Healing Traditions

NEW AFRICAN HISTORIES SERIES

Series editors: Jean Allman and Allen Isaacman

David William Cohen and E. S. Atieno Odhiambo,
*The Risks of Knowledge: Investigations into the Death of
the Hon. Minister John Robert Ouko in Kenya, 1990*

Belinda Bozzoli, *Theatres of Struggle and the End of
Apartheid*

Gary Kynoch, *We Are Fighting the World: A History of
Marashea Gangs in South Africa, 1947–1999*

Stephanie Newell, *The Forger's Tale: The Search for Odeziaku*

Jacob A. Tropp, *Natures of Colonial Change: Environmental
Relations in the Making of the Transkei*

Jan Bender Shetler, *Imagining Serengeti: A History of
Landscape Memory in Tanzania from Earliest Times to
the Present*

Cheikh Anta Babou, *Fighting the Greater Jihad: Amadu
Bamba and the Founding of the Muridiyya in Senegal,
1853–1913*

Marc Epprecht, *Heterosexual Africa? The History of an Idea
from the Age of Exploration to the Age of AIDS*

Marissa J. Moorman, *Intonations: A Social History of Music
and Nation in Luanda, Angola, from 1945 to Recent Times*

Karen E. Flint, *Healing Traditions: African Medicine,
Cultural Exchange, and Competition in South Africa,
1820–1948*

Healing Traditions

African Medicine, Cultural Exchange, and
Competition in South Africa, 1820–1948

∽

Karen E. Flint

OHIO UNIVERSITY PRESS
ATHENS

UNIVERSITY OF KwaZulu-Natal PRESS

Ohio University Press, Athens, Ohio 45701
www.ohioswallow.com
© 2008 by Ohio University Press
All rights reserved

Published in 2008 in South Africa by
University of KwaZulu-Natal Press
Private Bag X01, Scottsville, 3209
South Africa
Email: books@ukzn.ac.za
www.uknpress.co.za

University of KwaZulu-Natal Press ISBN 978-1-86914-170-7

Printed in the United States of America
Ohio University Press books are printed on acid-free paper ⊗ ™

15 14 13 12 11 10 09 08   5 4 3 2 1

Portions of chapter 4 appeared previously in the author's "Competition, Race and
Professionalization: African Healers and White Medical Practitioners in Natal,
South Africa in the Early Twentieth Century," *Social History of Medicine* 14, no. 2
(August 2001). Portions of chapter 5 appeared previously in the author's "Indian-
African Encounters: Polyculturalism and African Therapeutics in Natal, South
Africa, 1860–1948," *Journal of Southern African Studies* 32, no. 2 (June 2006).

Library of Congress Cataloging-in-Publication Data
Flint, Karen Elizabeth, 1968–
Healing traditions : African medicine, cultural exchange, and competition in
South Africa, 1820–1948 / Karen E. Flint.
    p. ; cm. — (New African histories series)
Includes bibliographical references and index.
ISBN 978-0-8214-1849-9 (hc : alk. paper) — ISBN 978-0-8214-1850-5 (pbk. : alk. paper)
1. Medicine—South Africa—History. 2. Traditional medicine—South Africa—
History. I. Title. II. Title: African medicine, cultural exchange, and competition in
South Africa, 1820–1948. III. Series.
    [DNLM: 1. Medicine, African Traditional—history—South Africa. 2. Colonialism—
history—South Africa. 3. History, 19th Century—South Africa. 4. History, 20th
Century—South Africa. 5. Missions and Missionaries—history—South Africa.
6. Witchcraft—history—South Africa. WZ 80.5.B5 F624h 2008]
    RA418.3.S6F65 2008
    615.8'80968—dc22

2008036486

Contents

Illustrations

Preface

This book, which is divided into two parts, begins in the 1820s. This marks both the early years of the Zulu kingdom (1820–79) and sustained interaction between African healers, white traders, and missionaries. The first part of the book examines changes in the medical, social, and political role of healers in the Zulu kingdom, particularly as Zulu kings and chiefs sought political consolidation. The second part investigates how local ideas of health and healing changed under white rule, first in Natal and later in Zululand after the Zulu kingdom's defeat in 1879. Intercultural encounters play an important role in both sections of the book, though in very different ways. In part 1, an examination of healers and healing in the Zulu kingdom not only sets up a basis from which to observe historical changes in local therapeutics, but explains why white observers of the period advocated specific strategies to tackle the social and political power of healers and witchcraft allegations within Natal. The second part examines a much wider array of cultural encounters, particularly as Zulu-speaking healers and patients encountered white and Indian populations and coped with the implementation of white rule. I end the investigation in 1948, which marks the beginning of the apartheid era as well as the decline of African healing associations and their struggle for legal recognition within Natal and Zululand. While certain continuities from the period under investigation persisted during the apartheid era—for instance, Natal and Zululand both licensed *inyangas* (traditional herbalists) up through the 1980s—apartheid also brought about a much greater sense of separation between population groups, making it easier to forget the cultural fluidity and contestations of these earlier times.

"Healing the Body: Disease, Knowledge, and Medical Practices in the Zulu Kingdom" (chapter 1) demonstrates that African healers healed individual bodies through pharmacological and surgical interventions as well as ancestral interference and unveiling of witches. Consequently, this chapter, which provides a description of the basic ideas of health and well-being in the Zulu kingdom, corrects earlier and often cited works that paint this period as static and cast African therapeutics largely under the rubric of superstition. The period of the Zulu kingdom not only brought together healers and remedies from throughout the kingdom, but provided unique challenges as new diseases and epizootics were introduced by its colonial neighbors to the west and east. Understanding cultural and medical approaches to the body and its ailments during this period helps explain why certain biomedical drugs and procedures such as inoculation later came to be adopted or sought after while others such as amputation or even pills were rejected. Indeed much of the specialized knowledge of herbs, gathering techniques, and medical practices of the nineteenth century changed greatly with the rise of urbanization, migrant labor, and a consumer culture that offered general remedies for a general public.

The next chapter, "Healing the Body Politic: *Muthi*, Healers, and Nation Building in the Zulu Kingdom," argues that healers not only maintained the corporal body but played an important role in maintaining the body of the nation. According to oral histories describing this period, the power of a ruler's *muthi* (African medicine) directly determined his or her success—both over political rivals and armies and in maintaining the favor of the community. This role placed certain healers in particularly close proximity to and relationships with Zulu kings and chiefs, which sometimes created untenable tensions. These healers helped maintain political rule by performing specific rituals that strengthened the nation and chiefdoms and by helping to point out those guilty of crimes and witchcraft. Healers who were disloyal or tried to overshadow a chief or king were seen as destabilizing and unacceptable. Understanding the close relationship between muthi, healers, and political power is crucial to grasping the reaction of white settlers and the colonial government to African healers and African therapeutics.

The third chapter, "Early African-White Encounters: Healers, Witchcraft, and Colonial Rule, 1830–91," examines this very reaction among the earliest white settlers and during the establishment of Natal. This chapter argues that muthi and African healers were perceived as both a direct and an indirect threat to British colonial rule and Christian missionary endeavors. Healers correctly surmised that the colonial administration and Christian missions sought their destruction, leading some to engage in competition that questioned the morality and practicality of Europeans. The main purpose of this chapter is to trace the history of encounters between white traders, settlers, and missionaries and African healers and African therapeutics during the nineteenth century. These encounters led to relationships that over time evolved from benign mutual appreciation to open and hostile conflict and finally uneasy resolution. Central to this conflict was the issue of witchcraft. African communities sought to discover and expose those who practiced it, while whites aimed to protect the accused and prosecute accusers. Real or not, accusations of witchcraft threatened in subtle and not so subtle ways to undermine British colonial rule. Healers' political role, in legitimating both African political power and the judicial system, led not only to the criminalization of one type of healer—the *isangoma* (diviner)—but to the legalization and licensing of another—the inyanga (herbalist). Nineteenth-century African women, deemed legal minors by colonial law and excluded from the cash economy, embraced and reinvented the practice of the isangoma. These women not only empowered themselves socially and economically, but also played a crucial role in maintaining and negotiating "local" ideas of health and healing throughout the nineteenth and twentieth centuries. This chapter thus shows how white rule influenced African medicine from an early period, particularly as colonists sought to reshape African ideas of witchcraft.

This interaction and influence, particularly between white and African medical practitioners, went both ways. In "Competition, Race and Professionalization: African Healers and White Medical Practitioners, 1891–1948," I trace the ways in which the licensing of inyangas, along with rural poverty, urbanization, and the rise of a consumer culture, radically altered the practice of traditional healers. The rise of successful and rich inyangas, combined with an overcrowded biomedical market, led to ideological and commercial competition between white biomedical practitioners and African healers in the early twentieth century. Such competition led to government legislation and a curtailing of inyanga practices. This chapter begins by examining the historical antecedents of this competition, particularly the commodification of African medicine, and focuses on the role that competition, race, and gender played in the construction of local biomedical as well as African ideas of medical authority. White women featured prominently in the discourse surrounding biomedicine's professionalization in South Africa. Advocates of biomedicine appealed to white fears of racial degeneracy by citing African doctors' and healers' attendance on white women and the real and imagined sexual, social, and political implications of such encounters. This chapter aims not only to document historical changes in African therapeutics in the early twentieth century but to complicate current ideas of biomedicine's colonial hegemony. Furthermore, it demonstrates how white biomedical practitioners with the assistance of white legislators helped to shape traditional African therapeutics through legal and legislative means.

In the final chapter, "African-Indian Encounters and Their Influence on African Therapeutics, 1860–1948," I argue that African-Indian medical encounters, combined with the pressure of biomedical scrutiny, helped to define and shape what is today considered traditional African therapeutics. Indians not only adopted the ailments of their African counterparts, but became practitioners and purveyors of African medicine itself. Such persons helped to introduce to the African population Indian herbs and substances that later became incorporated into the African pharmacopoeia. This is in contrast to "white" substances and remedies, which remained largely excluded. This chapter thus further explores how and why certain cultural ideas and practices with regard to medical therapeutics were adopted and incorporated, while others were rejected. By examining the historical antecedents of African-Indian encounters and African therapeutics in this area, I again demonstrate that what is considered "African" or "indigenous" knowledge is rather an amalgam of many cultural and political influences.

While a lot has changed since the late 1940s, healers in a postapartheid South Africa faced many of the same issues as their early twentieth-century predecessors—gaining government recognition, establishing medical authority, and protecting their profession from biomedical prospecting. I touch on some of these issues briefly in the epilogue. Legal recognition has, however, brought up more interesting issues regarding future regulation than I can do justice to here; they must await future study.

Acknowledgments

This book has been in the making for a long time, and thus I am indebted to many people and institutions who have helped contribute to its final completion. Much of the original research was conducted in South Africa during 1995 and 1998, with follow-up trips in 2002 and 2005. This project grew out of my experiences living and studying in 1992 at the University of Cape Town, where I had the fortune of working on an oral history project with Lance Van Sittert and Helen Bradford. These two scholars skillfully introduced me to the methodology of oral history and encouraged me to pursue my own interests at an early stage of my career. It was while investigating, under Dr. Bradford's supervision, the experiences of women who had had illegal abortions in South Africa that I initially discovered the important roles that African healers played in people's lives.

I am most grateful for the support of friends, teachers, and colleagues over the years who were instrumental in the inception, growth, and final production of this book and who provided much intellectual stimulus, emotional support, and, at times, necessary distractions. Without them this book could not have been written.

In its first incarnation as a dissertation and later in its expansion and refining as a manuscript, I benefited tremendously from the intellectual insight and probing questions of Ned Alpers, Cynthia Brantley, Catherine Burns, Colin Fisher, Doug Haynes, David Johnson, Oscar Lansen, Michael Mahoney, Boniface Obichere, Robert Papini, Julie Parle, Arthur Rubel, John David Smith, Yuki Terazawa, Sharon Traweek, Lance Van Sittert, George Vilakati, and William Worger. Thank you.

In South Africa, I am most grateful for friends who provided their support, hospitality, and at times interventions during my extended stays in that country. Thank you, Paula Despins, Robin Fable, Simona Gallo, Hein Kleinbooi, Khutso Mampeule, Pierre Matungul, Robert Papini, Julie Parle, Elphy Sibeko, Swazi Tshabalala, and Janet Wojcicki. During my research trips to the field and archives, several people helped to bring this project to fruition. I am especially thankful to Leo Naidoo, who not only welcomed me into his *muthi* shop, shared with me his love of the *muthi* trade, and introduced me to other healers, but became a good friend. V. T. Mkize, Environmental Health Officer of the Durban Health Department, was instrumental in introducing me to healers in the Durban area. I thank Anne Hutchings and Nicky Drysdale for opening their homes to me and for facilitating meetings with different healers. Paula Despins graciously shared her research data and site with me, introducing me to the rural area of Hlabisa and many of the healers who lived there. I am most grateful to the staff at the Pietermaritzburg Archive Repository and the Killie Campbell Africana Studies library where I spent a good deal of time. Not only were Pieter Nel, Unnay Narrine, and Thami Ndlovu in Pietermaritzburg and Bobby Eldredge and Dingane Mthethwa in Durban helpful, but their smiles and conversation provided welcome companionship that extended far beyond archival support.

I have been fortunate to have participated in so many supportive and intellectually thought-provoking seminars and study groups over the years. Many of the chapters in this book benefited from the constructive criticism and probing of seminar participants at the Pietermaritzburg and Durban campuses at the University of KwaZulu-Natal; the University of California, Los Angeles; the Anthropology of Science, Technology, and Medicine Seminar at the California Institute of Technology; and the history department at the University of North Carolina, Charlotte. In particular, I thank Catherine Burns, Keith Breckenridge, Karen Cox, Mary Dillard, Melissa Feinberg, John Flower, Cymone Fourshey, Christine Haynes, Jeff Guy, John Laband, Marianne de Laet, Michael Mahoney, Patrick Malloy, Tim Nuttall, Mel Page, Sonia Seeman, Shobana Shankar, John Smail, and John Wright.

I am particularly grateful to the many readers and editors at Ohio University Press. The critical insight of Catherine Burns, Allen Isaacman, and Jean Allman forced me to look at and expand this work in new and different ways. I especially thank Allen Isaacman, Jean Allman, and Gillian Berchowitz for their patience and dogged persistence as I completed the final manuscript.

The research for this manuscript was funded in part by the University of California, Los Angeles; a Fulbright Hays Fellowship; the American Institute of the History of Pharmacy; and several grants from the University of North Carolina, Charlotte. I am most grateful for the support of these institutions and their interest in the history of health and healing in Africa. Finally I would like to thank my family for their support over the years, and particularly David Johnson, who traveled much of this journey with me. Thank you for your many hours of listening to, reading, and editing my ideas in what has seemed at times like a never-ending process. Your love and support have meant so much to me.

Introduction

What Is "Traditional" about Traditional Healers and Medicines?

MR. MAFAVUKE NGCOBO AND THE CASE OF THE NOT-SO-TRADITIONAL HEALER

Mafavuke Ngcobo, a licensed African *inyanga* or "traditional" herbalist (as opposed to diviner or rainmaker) in the province of Natal, South Africa, gained the attention of white chemists and government authorities when he turned his small herbal practice in decidedly "untraditional" directions in the 1930s. In contrast to the colonial stereotype of the "witch doctor" reciting incantations to the dead over a mysterious bubbling brew, Mr. Ngcobo practiced medicine in ways uncomfortably familiar to white urbanites. Not only did Ngcobo use the title "doctor" (until reprimanded by the court), but he advertised himself in pamphlets as a "native medical scientist." Reverend Qandiyane Cele, who wrote a supporting letter to the Chief Native Commissioner, emphasized Ngcobo's credibility by pointing out that his "chemist" shops operated along "a European system."[1] By 1934, Ngcobo owned five *muthi* (African medicine) shops in and around the city of Durban, as well as a lucrative mail-order business that sold bottled medicinal remedies labeled in both Zulu and English. His remedies contained local herbs, some Indian remedies, and, more controversially, chemist drugs and patent medicines. In addition, in 1928 Ngcobo helped establish the Natal Native Medical Association, a professional organization of African herbalists that tested dues-paying members on their knowledge of "native curatives." This organization (albeit a small one) lobbied the government for "native medical rights," hired lawyers to defend its members in court, and became quite adept at capturing media attention. Such

1

seemingly "untraditional" behavior troubled the local white medical establishment, and in 1940 Ngcobo went on trial for "carrying on the business of a chemist or druggist."[2]

"Traditional" healers, of which there were many types, specialized according to their talents and calling. Historically they performed a variety of functions for African communities; these included bringing rain, detecting witches and criminals, "doctoring" armies, negotiating with ancestors, and using herbs and surgical procedures to cure and mend the body. Ngcobo's very success as a "not-so-traditional" healer, however, represented larger transformations in African healing practices that occurred between the 1820s and 1940s, particularly in the areas of Natal and Zululand (the former Zulu kingdom). This period saw healers transform themselves from politically powerful men and women who threatened to undermine colonial rule and law in nineteenth-century Natal into successful venture capitalists who competed for turf and patients with white biomedical (Western and allopathic) doctors and pharmacists in the early twentieth century.[3] Healers not only adjusted to the political, social, and economic factors that accompanied British colonial rule, but found their status dramatically affected by provincial legislation. Beginning in the 1860s, white legislators criminalized all types of healers, but in 1891 made the unique decision to license African midwives and inyangas. Not all traditional healers adopted Ngcobo's practices or competed with whites in the same manner, yet in a budding multitherapeutic society many rural and urban healers had begun to incorporate and experiment with medical and non-medical substances associated with South Africa's other population groups. This blurring of medical as well as cultural boundaries made white medical practitioners and government authorities quite uncomfortable and raised important questions regarding the very nature of so-called traditional medicine and the role of licensed inyangas.

Ngcobo's trial not only brought many of these particular issues to light but also demonstrated the difficulty in trying to characterize exactly what was "traditional" about "traditional medicine." On one side, white administrators, doctors, and chemists argued that "native medicines" were static and unchanging and should be defined largely as the absence of what was considered "white," that is, exclusive rights to biomedical ingredients, titles, tools, practices, scientific methods, and white patients. Conversely, the Natal Native Medical Association, Ngcobo, and his lawyer argued that African therapeutics were dynamic and experimental and changed with the times. Problems encountered in legally codifying this medical tradition, however, resulted largely from the ambiguity of past legislation, particularly the 1891 Natal Native Code, which had originally legalized and licensed the practices of African midwives and inyangas. The code's writers presumed the concepts of native and European medicines as self-evident and obviously and inherently different. The only legal restriction placed on licensed inyangas was a caveat specifically banning the sale of "love philtres or charms." In 1932, administrators sought to more clearly distinguish these two medical cultures by amending the code to explicitly state that inyangas "may prescribe, deal in and sell *native medicines only*."[4] The Ngcobo trial was yet another attempt to disrupt the "development of a new *hybrid* system" that had clearly begun at a much earlier date.[5]

In determining what constituted "native medicines" or "traditional methods" of healing, the prosecution, like the white administrators before them, sought to establish the "authenticity" of so-called native customary law by turning to the requisite African "experts" or old African men. At the 1940 trial, the testimony of Ngcobo's own employees, and an elderly licensed inyanga from Port Shepstone who bore no relationship to Ngcobo or his business, provided the bulk of the prosecution's evidence. Their testimony showed that although Ngcobo's own workers characterized some medicines as "European" and others as "African," they disagreed on the origins and use of others. Ngcobo's employee of twelve years, George Mvuyane, told the court: "Everything you buy from a chemist shop is a European medicine and everything you go and dig for is a native medicine."[6] Gonzaga Qhobosheane, another worker of Ngcobo's, and Ndabakohliwe Kuzwayo, the seventy-year-old exemplary inyanga, argued that native medicines also included animal fats, skins and bones, and minerals. With regards to medicinal plants, these African witnesses distinguished "native medicines" by whether they grew wild in Natal and Zululand. By the 1940s this involved a number of exotic species to include ones prepared by local chemists such as jalap (Zulu: *jalambu*) from South America and male fern (Zulu: *nkomankoma*) from Europe and North America. While Africans designated wild exotic plants as "native," some native plants, such as croton seeds, referred to by Mvuyane as *nhla kwa zaseIndia*, which indicated an Indian influence, assumed exotic connections. The origins of other substances such as mercury (Zulu: *sigiti*) seemed to defy classification, perhaps because both Ayurvedic (Indian) and biomedical practitioners used it. Kuzwayo said he knew no "medicine men" in Port Shepstone who used it, while Qhobosheane claimed it was quite common all over the country.[7] Despite the evidence of African healers and the fact that wild exotic plants introduced by early white traders and settlers had been growing in the area for more than a hundred years, the prosecution argued that "indigenous" medicine should include only *indigenous* plants of South Africa. The defense argued against a double standard that enabled the British pharmacopoeia to include many foreign herbs and materials but limited "native medicines" to indigenous South

African substances.[8] Instead the defense suggested a definition that included all herbs grown in South Africa and medicines regularly and long used by "native medicine men," such as jalap, male fern, and mercury.

The importance for the prosecution to clarify this first issue becomes evident when considering the next: how should licensed African practitioners secure and prepare their remedies? Again Kuzwayo, serving as the primary witness, testified that he dug up many of his own herbs and roots and killed snakes and iguanas, but bought fish bones from Indians and some "native medicines" in Durban (most likely from the native markets along Victoria Street). He did not purchase medicines from white chemists. While he agreed that inyangas typically used bottles for prepared medicines, he claimed that Ngcobo's shops had a greater number of bottles containing a larger variety of substances than he had ever seen in an inyanga's practice.[9] Native medicines, he testified, could be prepared by chopping, cutting, burning substances to ash, by cooking and by adding medicinal material to water.[10] Ngcobo's defense did not challenge these contentions, but again highlighted the duplicity of the law that enabled innovation among white medical practitioners while denying it to Africans. For instance, he pointed out that chemists had historically compounded their own medicine, yet the rise of wholesalers had made this practice less necessary. Certain herbal remedies regularly used by African herbalists, like male fern, were now available in tincture form (at many chemist shops). Why, the defense asked, could the African practitioner not also save time and energy through their purchase?[11] Given that Mr. Ngcobo was found to possess his own supply of various preservatives and patented medicines, these particular arguments—while raising important questions—came across as somewhat disingenuous.

In the end Ngcobo was fined £25 and charged with "not acting within the rights conferred upon him by his license as a native medicine man and herbalist."[12] The presiding judge rejected the defense's argument regarding healers' uses of nonindigenous substances. Instead he ruled that "native medicines" are "characteristically *native both in origin and composition,* that is medicines compounded and prepared from roots, bark, herbs, leaves, fats, skins and bones and other *indigenous* substances."[13] On appeal, Judge Feetham argued that this category should include "medicines such as natives can make for themselves by comparatively simple processes, not requiring a high degree of scientific skill, *out of the natural substances of the country which are available to them.*"[14] While it is not entirely clear from this latter characterization whether exotic jalap could be used by inyangas, the distinguishing feature for Feetham was the degree of sophistication with which these different medical practitioners mixed or compounded medicines. Like other imperial thinkers of the time, Feetham's emphasis reflected an imagined binary between European and African societies, the assumption of European rationality and science as contrasted to African irrationality and simplicity. Likewise, European society was deemed modern and innovative while African customs or traditions signified the past. African "traditions" were not meant to be adaptable or improved on but a temporary juncture that according to the judge would eventually give way "by degrees, as education and civilization extended."[15]

This court case is particularly interesting because it not only shows the difficulty in determining what is "traditional" about "traditional" medicine and healers in the 1940s, but also exposes some of the politics behind the courts' desire to codify African medicine. While Ngcobo's case provides evidence of medical cultural exchange, it also demonstrates the establishment's decided lack of interest in certain group interactions—such as those of Africans and Indians—which are evident but unremarkable to the court.[16] Instead legal arguments were cloaked under the veil of maintaining African "authenticity," while the real contention remained one over economic and ideological competition between African and white medical practitioners. This competition, rarely acknowledged, stretched far back into the nineteenth century and involved a variety of actors. Such rivalry is essential to note, as it not only upsets conventional notions of traditional African and biomedical medicine but demonstrates that medicine was yet another arena for larger colonial contests over political and cultural hegemony. The results of this competition influenced the ways in which biomedical and African healers came to conceive of themselves and largely limited healers' legal status under white rule.

Some sixty years later, in August 2004, a multiracial and democratically elected South African parliament officially recognized traditional healers and set in motion a legal framework to license them throughout the country. An Interim Traditional Health Practitioners Council was charged to "provide for a regulatory framework to ensure the efficacy, safety and quality of traditional health care services; to provide for the management and control over the registration, training, and conduct of practitioners, students and specified categories in the traditional health practitioners profession; and to provide for matters connected therewith." Consequently, more than 350,000 traditional healers, who attend the majority of the South African population, will gain access to the benefits and burdens of medical regulation. While the recently adopted Traditional Health Practitioners Bill is generally more descriptive than past legislation, the legal definitions used to characterize traditional medicines and health practitioners are still somewhat ambiguous. Ironically, like earlier medical legislation under white rule, the 2004 bill defines traditional medicines and practitioners by the absence of biomedical substances and practices and assumes them to be self-evident. For instance, the law stipulates that "traditional health practices" are based on "traditional philosophy," which is then defined as "indigenous African techniques, principles, theories, ideologies, beliefs, opinions and customs and uses of traditional medicines . . . which are generally used in traditional health practice."[17] Again, while such a definition might seem straightforward and obvious to most, the terms "traditional" and "indigenous" mask a complicated history and more problematically pose the assumption of "indigenous" traditional medicines.

In this book I explore the history of what is today deemed "traditional" African medicine. In particular, I examine how the practice of African therapeutics and whites' perceptions of African healers changed, both during the precolonial period and as traditional healers faced the challenges that came with white rule. In doing so, I analyze two related phenomena: how knowledge and culture are used to assert, challenge, or elide hegemonic forces and how such notions are adapted, produced, and negotiated between population groups with different interests, cultural beliefs, and access to power. My objectives are twofold: (1) to show that while African or local medicine has maintained certain core beliefs over time, it also has been dynamic and sometimes open to non-African beliefs, practices, practitioners, and substances; and (2) to demonstrate that within African history, medicine was an important site of power, contestation, and cultural exchange that not only reflected but also affected intergroup relations. By indicating the fluidity of African therapeutics and the dynamics of group encounters regarding health and well-being, I seek to challenge conventional understandings of cultural and medical boundaries. This study is grounded in the specific historical circumstances of South Africa, predominantly within the culturally plural province now known as KwaZulu-Natal. By beginning with the founding of the Zulu Kingdom in the 1820s and ending in 1948 with the election of the Nationalist Party, I examine the role and history of African medicine and healers in an independent African nation, during the period of colonial encroachment and white rule, and before the onset of apartheid rule. Though some of the phenomena discussed in this study are unique to the province of Natal and Zululand, like the licensing of healers and its large Indian population, the province's history helps to shed light on the transformations and dynamics of local therapeutics and intergroup encounters found in many colonized states.

Of necessity, this study moves away from more conventional medical histories and notions of medicine, particularly in its consideration of medicine's power to affect social and political change. Beginning with the emergence of the Zulu kingdom (1820–79), it is clear that African healers played an important role in upholding the authority and national identity of a new Zulu nation.

African healers and therapeutics healed not only the physical and social body, but the body politic. As the Zulu kingdom emerged and consolidated its power, medicine helped defeat Zulu enemies, strengthen the king, and create a new sense of national pride and obligation (see chapter 2). Even after the defeat of the Zulu nation by the British in 1879, healers continued to play an important role in maintaining local beliefs and power structures that challenged British rule. Some of the most contested moments of the colonial encounter occurred when African communities suspected witchcraft as the cause of death or illness and defied British law by seeking out local healers to confirm their suspicions. By expanding this study to include wider and local notions of medicine, we see how African therapeutics were used as a form of social control and a tool of the Zulu empire, but also why white native administrators and legislators in neighboring Natal (1830–91) reacted as they did to a certain subset of African healers within their own territories (see chapter 3).

To understand how people conceive what is "traditional" about traditional medicine, we must consider cultural actors and processes not commonly associated with African therapeutics—that is, white biomedical practitioners, Indian healers, and the implementing of white rule. As the Ngcobo case demonstrates, within KwaZulu-Natal, important players included African healers and their patients, indentured and ex-indentured Indians, white biomedical practitioners, and government administrators. White rule contributed both deliberately and accidentally to the rise of a multitherapeutic society as well as to the interaction of these various parties. While the region's medical "traditions" and cultures—African, Indian, and European—often have been treated as their own "systems," bounded and separate from each other, this book argues that this was not the case.[18] Not only did each community exhibit its own medical plurality, but the historic interactions of these various cultural entities affected the practice of the others to varying degrees. Understanding the history of medicine in South Africa thus requires an examination of how these groups interacted and the sites, actors, and circumstances that constructed these medical cultures.

Medical competition between the region's various therapeutic groups played an important role in this interaction, in turn influencing how South Africa's different medical practitioners came to envision themselves and their own medical authority. Again, as the Ngcobo case illustrates, biomedicine sought to establish its authority by invoking notions of science and racial superiority while legally restricting "African" medicine and its practitioners from using "white medicine" (see chapter 4). African and Indian practitioners, on the other hand, sought to modernize and professionalize their occupations by winning the confidence of a new multiracial urban clientele while

Who gets to represent traditional healers? Traditional healers posing in the 1880s and in 2002.

Henry Kisch, *Photographs of Natal and Zululand Album* (1882). Killie Campbell Africana Library, C8105.

Indian *inyangas* in an Indian *muthi* shop in Ladysmith, KwaZulu-Natal. Photo by author, 2002.

also seeking to circumvent the legal restrictions imposed upon them. An examination of the political, social, and economic encounters of such practitioners and their patients demonstrates the multicultural origins of so-called indigenous medicine and the ways in which African medicine negotiated and sometimes resisted its encounters with South Africa's other medical communities. By delineating the plural cultural heritage of African medicine, I seek to raise important questions regarding traditional medicine's "indigenous" nature, demonstrating that groups such as Indian inyangas not only used so-called indigenous medical knowledge, but shaped and contributed to it as well (see chapter 5). Consequently, my work challenges academic and popular notions of cultural exchange in South Africa and contributes to a growing body of scholarship focused on the construction of cultural identity.

EXAMINING NOTIONS OF TRADITION

Tradition has been a central concern to scholars of Africa since the early twentieth century, though understandings of this somewhat nebulous concept have changed radically since then. Initially the domain of anthropologists and colonial administrators, studies of African tradition helped colonists to draw up customary law and better implement colonial rule. Such scholars were often motivated by the belief that tradition needed to be documented and preserved against the rising and destructive surge of modernization, colonialism, and urbanization. Consequently such studies often ignored the rapid transformation that accompanied white rule, seeking instead to describe African cultures as free from European influence and as largely static and unchanged. E. E. Evans-Pritchard's seminal work on witchcraft and the Azande, for instance, unrealistically insisted that despite the Azande peoples' having been displaced and forced into government settlements set up to control sleeping sickness, such events had "not produced any great change in the life of the Azande."[19] The search for African "authenticity" and tradition was largely replaced in the 1940s by studies that examined the very transformations and structural changes engendered by white rule while still treating what preceded it as static.

By the 1980s, there seemed to be a major rethinking of tradition, in terms of both what it was and how it was used. Jan Vansina wrote influentially about the meanings and history behind African oral traditions—the stories passed down through generations that usually told of past rulers and the establishment of a people or a nation. Some traditions, such as that of "Sundiata," which told of the establishment of the Mali kingdom, went as far back as the thirteenth century. Vansina argued such traditions reflected the present as much as they offered a window into the past. He demonstrated how performers appealed to the needs, questions, and desires of a contemporary audience.[20] While Vansina believed that it was possible to access evidence of the past through such traditions, particularly through examining metaphor and comparing regional oral traditions, it was also clear that new historical circumstances meant such traditions had often been changed by various narrators over the years. Such adaptability, Vansina argued, enabled the very durability and relevance of traditions. As he wrote later, traditions "must change to remain alive."[21] Yet he also warned that traditions could perish altogether when the basic principles underlying a culture or society were forgotten or discarded in favor of another incongruent tradition.[22]

Adopting this idea of a changing and adaptable tradition, other scholars sought to examine the ways in which tradition had been constructed. Following the influence of Terence Ranger and Eric Hobsbawm's *Invention of Tradition* (1983), African historians began to question the very nature of what had been construed as African "tradition." Clearly European discourse on African "tradition" had been self-serving, leading white rulers such as the judge presiding over Ngcobo's court case to create a false binary that painted African "tradition" as the antithesis to European "modernity." Yet what impact did such a discourse, and, as V. Y. Mudimbe argues, an academic scholarship on Africa replete with non-African categories and epistemologies, have on African practices?[23] Scholars turned to examining the ways in which colonialism and customary law helped to construct rather than reflect so-called traditional African identities, customs, ethnicity, religion, and gender relations. Authors argued that colonial officials with limited knowledge of African realities had codified, classified, and changed African cultures. In doing so they had helped to shape and influence that seeming reality. Ten years after publishing his and Hobsbawm's seminal book, Ranger reexamined how his initial thesis had been expanded and challenged by more recent work on tradition and Africa. He concluded that it was necessary to revisit some of his original assumptions about the power of the colonial state to shape tradition. Likewise, scholars began to ask how successful colonists had been in reifying these so-called traditions. While the colonial state seemingly changed and shaped tradition in the public arena, research showed that it had less of an impact in the private sphere. Ultimately colonists could not control the ways in which practices and ideas were interpreted or imagined by the colonized.[24] While earlier scholars had posited a break between precolonial and colonial traditions, later historians challenged this.[25] Steven Feierman, for instance, agreed that colonialism had been interventionist but pointed to African agency in the shaping of new traditions. In particular he examined the role of "peasant intellectuals" who used both the past and an enlarged "tradition" to meet contemporary needs.[26]

More recently, scholars have expanded their examinations to the *process* by which tradition was/is constructed. This includes studying the role of African mediators, often male, literate, politically invested, and with access to white administrators. Such actors served as cultural brokers or translators who not only helped to make colonial ideas and practices more palatable to Africans but also had an influence on the way colonial administrators understood and reacted to the African community.[27] In essence, tradition, like many other concepts, is not only mutable but can be employed strategically for specific ends by both the colonized and the colonizer.

Scholarship on African therapeutics has in many ways mirrored the historiography of tradition, primarily because African therapeutics itself has been identified as an important component of African tradition. Yet attention to its historical construction and the importance of its contributing actors and sites has been largely absent. By examining African therapeutics historically, I show that traditions are fluid, contested, open to cultural exchange, and a means of asserting power. Furthermore its construction involves both colonial meddling and the work of cultural brokers and also reflects the will and concerns of the general public.

The question of tradition and therapeutics became remarkably important in the years just prior to South Africa's first democratic elections. A rash of human muthi murders and mutilations as well as brutal killings of alleged witches gained the attention of the national press. Healers who were by association implicated in such practices used notions of tradition to counter the negative publicity. Obtaining human muthi, which usually involved the removal of certain body parts from a live person, was not executed by a healer but carried out at his or her instruction.[28] From these body parts healers concocted powerful medicines alleged to enhance the wealth and power of the recipient. Likewise the killing of alleged witches, whose prevalence in the northern provinces led to the establishment of two government commissions (1995, 1998), often depended on traditional healers to "smell-out" or identify so-called witches, who were then targeted by vigilante youths.[29] Although such incidents have a historical precedent in places such as the Zulu kingdom, where African healers did advocate the use of human muthi and named witches, in the past these practices had taken place under radically different circumstances and, at least theoretically, had not been used for personal gain. The publicity surrounding the violent incidents in the 1990s not only implicated traditional healers both directly and indirectly but tarnished the very image that certain healers had been working so hard to promote. From the 1980s on, healers seeking national recognition for their occupation attempted to reassure national legislators and the public of their honorability. In the early 1990s, traditional healers launched a public relations campaign that actively condemned muthi murders and the killing of alleged witches as "untraditional" and the work of charlatans rather than genuine healers.[30] Politically savvy traditional healers thus employed notions of tradition to distance themselves from and restrict what they considered antisocial behavior while legitimating their own practices.

Today South African healers use ideas of tradition to emphasize their authenticity and legitimacy in a multicultural environment where patients may choose from a variety of different practitioners and therapeutics. Healers thus refer to themselves in English as "traditional healers" and argue that their practices have not changed over time but reflect the practices of their forebears. This reference to ancestors alludes not only to the knowledge passed down through the generations but to the active role that ancestors play in the therapeutic process and passage of knowledge through dreams, trances, and visions. The term "traditional healer" also avoids the negative connotations and inaccuracy of "witch doctor" and is less exclusive than the colonial-derived terms "medicine men and herbalists." Instead, the designation encompasses both women, who make up a large percentage of healers, and other types of healers besides herbalists.[31] This term is used interchangeably but consciously throughout this book with the term "healer" or "medical practitioner."

The utility of tradition thus emerges from its flexibility, the way in which it can be called on to support or condemn various actions, practices, and/or beliefs. It is a catch-all concept that connotes the passage of items, images, symbols, events, beliefs, behaviors, customs, or practices from one group or generation to another over the years. When people accept and acknowledge tradition, it is usually because they see its worth and value its continuity. On the other hand, those who find a tradition disagreeable may cast aspersions, claiming tradition to be outdated, nonsense, or superstition. Either way, traditions have tremendous power in their ability to bind people together or to cause generational, family, or community strife and strain. In a postapartheid South Africa, many are reasserting the relevance of "African tradition" and rediscovering "traditions" long disparaged by the previous colonial and apartheid governments. Others, however, are challenging "traditions" such as public virginity testing that have been deemed harmful and discriminatory to girls.[32]

Sociologists have determined that the concept of tradition often gains importance during periods of rapid social, economic, and political change; it becomes a means by which to assert a feeling of power over events that maybe outside of one's immediate control.[33] This seemed to be the case when I began this project in South Africa in 1998. As a historian, I was particularly interested to understand how practices, beliefs, and values of healers had changed over time and space. Yet when I conducted interviews with healers in KwaZulu-Natal, I discovered that they often did not share this concern, and some vehemently rejected this. Instead traditional healers expressed an interest in talking about what practices and forms of knowledge had remained the same; they sought to write down "the tradition" before it was forgotten. They feared that African youth, in the face of larger social and economic changes, did not value or understand their own history and traditions and would soon lose them altogether. In the face of increased urbanization and globalization and a disappearing local flora and fauna, healers were/are not alone in this fear. During the past ten years, a number of government and university research projects have sought to record the botanical knowledge of South African healers and create a comprehensive database of South African plants for these very same reasons.[34] When asked why he worked and shared plant remedies with the pharmacology department at the University of Cape Town, healer Phillip Kubekeli replied, "There must be something written for our future generations, if there is nothing written all our knowledge will just collapse. Truly, look at our children . . . our children don't even know the first principle of our primary healthcare."[35]

Tradition, however, is also often contested as people inevitably have conflicting views over its meaning, practice, or use. As in the Ncgobo trial, Zulu-speaking healers with whom I spoke in the late 1990s often had competing notions of what constituted "traditional" medical practices—be it the appropriate gender of practitioners, how to collect and mix medicines, or the means of acquiring patients. Recently, healers in KwaZulu-Natal were debating the legitimacy of processed herbs and whether this practice falls within the canon of "tradition."[36] The processing of herbs has been practiced for the past century and is largely but not entirely an urban phenomenon which continues to be practiced by a number of KwaZulu-Natal healers and muthi sellers. Yet who should decide whether this practice falls under the realm of "tradition"? What is the process of negotiating such determinations? And how far back does one need to go in history to define "authenticity"? Answering these questions, let alone making them a part of public policy, enforceable by law, is where "tradition" becomes a thorny and controversial issue.

Another important question to ask is, if traditions are so highly valued because of the strong emotive link they provide to the past, can traditions also change and modernize? With regard to the debate over processed herbs, some healers as well as botanists argue that the processing of herbs combined with sustainable harvesting can better preserve a rapidly depleting local flora and consequently the practice of traditional medicine.[37] Others point to the need for new forms of preparing herbs to meet the coming regulation and systematizing of traditional medicine planned by the South African government's

Medicines Control Council.[38] While it is unclear what African healers will decide on this particular issue of tradition, the question is clearly prompted by changes within African therapeutics in the past and present. In many ways, traditions are like family recipes; they are not static but subject to change over time. Each generation may improvise to accommodate the availability of ingredients or transformations in technology or to better suit their own tastes, all while attempting to honor the basic recipe. In essence, tradition is like an elastic fabric that is often stretched to meet specific needs. In this way we can see tradition as a cultural construct, subject to change from within and without. While there may be certain values, practices, and symbols that persist over time, there is nothing authentic or essential about them as their importance or meanings shift to reflect society's norms and values. This is not to say that they are meaningless. On the contrary, they have great importance to those who abide by them and possible repercussions for those who do not or for those who are intentionally excluded.

In order to understand what is "traditional" about "traditional" medicine and healers and what has been constructed as "traditional," I seek to answer some contemporary questions: Why, if certain local herbs are available in pill form, do patients and clients insist on more traditional preparations of these herbs? Do patients see this as primarily an issue of functionality, or does the ritual of preparation serve a deeper social and cultural purpose? Why are pills and many European substances rejected while Indian herbs are embraced by patients and their healers as "African" or "traditional" medicine? Why did the use of precollected herbs and already prepared herbal mixes become largely acceptable, while early to mid-nineteenth-century methods indicate that muthi should be collected and prepared only when a person is ill? Why do many South Africans continue to seek the support of traditional healers despite the fact that, until recently, they have not been covered by medical insurance and their services are often much more expensive than those of biomedical practitioners? Finally, are the current practices of traditional healers the result of historical interactions or merely a reflection of today's multitherapeutic market? By engaging these questions in a historical manner I hope to better ascertain how it is that "tradition" was and continues to be created or imagined and to understand how and why certain aspects of culture have been jettisoned, others maintained, and others—completely new—incorporated over time.

CULTURAL BOUNDARIES AND BROKERS

As a social-cultural historian I am interested to see how culture—in this case an African medical "tradition"—is constructed and reconstructed in the past;

Advertisements for toll-free numbers became legal in South Africa in 1992. Healers sought to sell their services to a modern public by appealing to notions of various "traditions." These ads appeared in the *Sowetan* in December 1992.

where and how cultural boundaries are drawn; the circumstances that prompt remapping; the role of cultural brokers who assert or assign these boundaries;[39] and the sites where boundaries are produced and reproduced.[40] This book looks particularly at the role that cultural brokers—African, white, and Indian— as well as patients played in shaping the boundaries of "African" therapeutics. Cultural boundaries are important as they determine cultural identity; who is included or excluded—who is Zulu or not Zulu, Indian or not Indian. And in the case of healing, they determine the type of healing—whether practices are identified as African, Indian, or European. Despite the labels applied by white administrators and the general public, these categories are seldom so discrete in reality. Yet notions of such differentiations between cultural groups gain importance when people perceive tangible social, economic, and or political privileges or disadvantages in belonging to one group versus another. This has been true in the case of healing as well.

Cultural boundaries, like tradition, shift to meet changing circumstances. This was particularly true in the rapidly changing and multicultural community of what is now KwaZulu-Natal. While many Zulu-speakers emphasized clan identity in the nineteenth century, a more widespread Zulu identity emerged in the twentieth century as a result of white imperialism and growing Zulu cultural nationalism.[41] Likewise, the South African Indian identity today encompasses a heterogeneous mix of people claiming South Asian, Middle Eastern, Zanzibari and/or Islamic heritage.[42] This overarching Indian identity emerged between the late nineteenth and mid-twentieth centuries due in part to its imposition by a white-dominated government, but also as communities asserted it to organize and defend themselves politically. Similarly, white South Africans—a number of whom have a mixed genealogy and heritage—may remember genealogical lines from various parts of Europe while forgetting the contributions of Khoisan and enslaved communities of the Cape. Despite the assertion or assignment of boundaries, however, groups are not homogenous, and individuals have multiple identities. As anywhere in the world, South African communities often vary in terms of religion, language, and regional origin and are further split by class, gender, generational, and urban/rural divides. What it means to be "Zulu," "white," "Indian," or, more recently, "African" in South Africa is imagined in many different ways over time and space.

Cultural boundaries are not only prone to shift over time but are porous by nature, particularly where and when cultural groups encounter each other with frequency. What passes as African, Indian, or European medicine has also changed over time. In other words, there is potential for leakage—the diffusion, adoption, and appropriation of other cultural ideas, practices, and artifacts. The result is a polycultural amalgam that blends together various strands of influence, creating new and sometimes unexpected patterns in the cultural fabric.[43] Historians Robin Kelly and Vijay Prashad argue that all cultures are essentially polycultural with a variety of entrance points.[44] Unfortunately, present-day antagonisms between various groups are seen by many in the general public as natural, the result of culturally distinct groups, rather than as cultural productions that arise from specific historical circumstances.

Within South Africa, white legislators worked hard to reinforce these notions of cultural and racial difference. British segregationists and apartheid legislators sought to hinder the shifting and porous nature of cultural boundaries through bureaucratization and segregation. Yet even under apartheid's Population Registration Act (1950), individuals could be racially and culturally reclassified.[45] This was sometimes done at the insistence of the state; at other times individuals asserted new identities for practical purposes. Muslim Indian businessmen in the Cape, for instance, sometimes sought reclassification as Malay (a cultural subset of the Coloured population) to maintain property and businesses in "Coloured areas" designated by the Group Areas Act (1950).[46] White South African legislators, however, focused primarily on preventing cultural exchange between "white" and "non-white" groups, even to the extent that apartheid segregated hospitals, and ambulances were forbidden to pick up patients of the "wrong" race.[47] Less vigilance was paid to the interactions among "non-white" groups, though legislators also sought to disrupt the permeable boundaries between these groups as well. Divide-and-rule tactics and ideologies of difference were reinforced in race-based apartheid-era schools, through the Group Areas Act, in the public media, and, not surprisingly, these ideas of difference continue to linger in today's public imagination.

In a postapartheid era it is again necessary to revisit these assumptions about cultural difference and to challenge popular notions of group relations. Viewing all cultures within South Africa as polycultural fusions of African, European, and Indian Ocean influences and appropriations allows one to complexify notions of culture and cultural exchange and demonstrate the historic interconnectivity of present-day cultures. But more importantly a history of long-forgotten encounters may disrupt normative ideas of the "cultural divide" that separates communities today. Delineating cultural threads within those exchanges, however, is not always an easy task, particularly as many ideas, practices, and artifacts have been incorporated to such an extent as to be seen as natural or "indigenous" to a particular group. This book attempts to delineate some of these threads with regard to the history of health and healing. In this way, this book serves as an empirical intervention to highlight these connections.

As a writer and historian it is sometimes difficult to avoid perpetuating conventional notions of tradition, culture, and race. The very appellations "European," "white," "Asian," "Indian," "African," and "Zulu" connote borders or flatten and homogenize what are otherwise quite diverse groups with different interests, histories, and changing identities. Qualifying a person as "Zulu," for instance, can inadvertently suggest a Zulu identity or support for the Zulu king, which was or is not always the case. This issue can be circumvented to a certain extent by utilizing the term "Zulu-speaker" or by referring to practices of the precolonial period as "within the Zulu kingdom." The kingdom's boundaries (though shifting over time) can be outlined either physically or by association. Within this text I often use the more general term "African" to refer to peoples indigenous to southern Africa; this is not to insinuate that the practices of "Africans" described herein can be attributed to all groups within Africa. Rather this is a way of acknowledging a more heterogeneous group of Africans. Natal became home not only to Africans who purposely escaped the Zulu kingdom, but to other non-Zulu-speaking Africans who sought work or better opportunities in Natal. Likewise the government made distinctions for licensing based on ideas of race—whether or not one was a "native"—rather than Zulu versus Sotho. While most African healers discussed in this book were Zulu-speakers, there were others, such as the wealthy and rather high profile Sotho healer Israel Alexander, who played a prominent role in the Natal Native Medical Association.

As researchers we are somewhat handicapped by the ways that previous writers have recorded history. Archival records, for instance, often refer to persons simply as "Indian" (sometimes "Arab" during the earlier period) and "native"; they do not always make subtler distinctions. Where possible I have noted individuals' and groups' class, religion, gender, generation, and level of education (to name but a few variables). Unfortunately both the history of South Africa and its sources lead me to use terms that I am simultaneously seeking to undermine and complicate. I hope by showing how "traditional" medicine in the Zulu kingdom and Natal is dynamic, adaptable, and polycultural, I can contribute to larger discussions about the ways in which a Zulu cultural identity has been constructed in the past as it is in the present.

CULTURE-BOUND ILLNESS AND MEDICAL PLURALISM

What makes South Africa particularly interesting for studying notions of tradition and health and healing is its multiculturalism. In the 1930s and '40s the motto of the South African Union Health department warned: "Disease Knows No Colour Bar."[48] Yet the interpretation, meaning, experience, and treatment of disease often varied greatly among South Africa's different ethnic and racial communities. This is because concepts of health, wellness, and the body, like tradition, are informed by our own experiences and the culture and era in which we live. Different communities often have various ways of understanding the body and illness and consequently diverse approaches to health and healing. For instance, biomedical conceptions of the body privilege a fairly mechanistic understanding of our selves, whereas Ayurveda (what is construed as "traditional" Indian medicine) views the body as consisting of five elements whose make-up determines one of three major body types. In southern Africa, a number of cultural-linguistic groups share the notion of an *inyoka* (an invisible internal snake) or force that resides largely but not only within the torso.[49] In addition to bodily understandings, many medical cultures consider the environment—physical, emotional, and spiritual—a potent force on well-being and thus a necessary component for determining the root of illness. Within Durban's heterogeneous Indian communities in the 1950s, for instance, illness could be attributed to retribution for sins in a former life, a visitation from a deity, losing faith in a Christian god, neglecting a Hindu house spirit, or witchcraft sent by either Indians or Africans. Determining the diagnosis enabled the patient to seek the proper practitioner and therapy, while an ambiguous diagnosis may have led to a more cautious approach in which several therapies were simultaneously pursued.[50] ·

Interpretations of wellness thus reflect society's larger cosmological ideas of how the world works and what is healthy and normal. For instance, male-pattern baldness and menopause, sometimes treated as "pathologies" within certain cultures, are readily accepted as natural parts of life in others that do not require medical attention. But culture can also cause individuals to experience very specific bodily symptoms and ailments that may be unique to that particular culture. This could be *umhayiso* or hysteria caused by the administering of love charms to Zulu-speaking women,[51] a trance entered by Hindu devotees during a Kavady festival, or illnesses such as anorexia and bulimia found in communities that promote thinness as an ideal. Anthropologists often refer to these experiences as culture-bound syndromes, as they are not generally shared by other cultural groups.[52] We may ask: To what extent do cultural practices, ideals, and interpretations influence our sense of well-being? Take for example the rise of reported persons suffering from attention deficit disorder in the United States during the 1990s. Did this reflect American society's decreasing tolerance for distracted or active children and adults? Was the disorder an outcome of the high demands and stress in a postindustrial capitalist society? Or was the record increase in biomedical doctors' diagnoses due to public demand after a new direct-to-public advertising campaign that popularized the symptoms and promoted a means of treatment?[53] Most likely, it was a combination of societal expectation, cultural demands, and patient awareness.

Within multicultural societies where multiple medical cultures survive side by side, people often search out practitioners who share their own medical ideas of the body and their etiology of illness. But many patients also utilize other practitioners and therapies in a plural medical culture. This may be because their own medical remedies failed or became unavailable or expensive, or because other therapies were perceived as more efficacious.[54] Others may use different therapies without knowing it, as with the case of healers who are trained in two or more medical cultures—an African herbalist who also uses the tenets and remedies of homeopathy. Another motivation, however, is that so-called culture-bound syndromes seem to cross cultural boundaries in medically plural societies—a Zulu-speaker becomes anorexic, or an Indian acquires the illness of an *umtwasa* (an illness and initiation undergone by certain African healers). In such cases, the appropriate cultural practitioner, the one most likely to affect a cure, is sought out. For instance, Africans in Natal historically preferred biomedical practitioners to treat syphilis,[55] a disease introduced and associated with whites, while Indians often sought out Zanzibari or African healers in cases of alleged witchcraft.[56]

Patients may also utilize other medical practitioners when compelled to do so by economic or legal necessity. The latter was certainly the case with indentured Indians on the sugar estates of Natal, African miners on the Reef, and, until recently, any South African worker hoping to claim medical aid or compensation. When medical pluralism is coerced through the dominance of one medical culture over another, as it was in South Africa, it can become problematic for several reasons. Although some medical concepts can resemble each other—for instance, Nguni notions of strong and weak blood are similar to biomedical ideas of the immune system—not all concepts are translatable.[57] This can lead to inappropriate treatment and frustration on the part of doctor and patient alike. The treatment may be ineffective or resisted if it does not fit within the cultural logic of the patient, for instance feeding newborns only breast milk seemed ludicrous to many African mothers at the turn of the century.[58] Likewise, this cultural gap has resulted in the current placement of persons experiencing symptoms of a *twasa*-initiate into mental institutions by biomedical doctors and patients' families who do not understand or recognize such culture-bound syndromes. When one medical culture (such as biomedicine) dominates and has legal privilege to enforce such decisions, as was seen in the Ngcobo case, other medical practitioners and therapies are by extension placed at a disadvantage in terms of their rights to practice and gain access to state resources. Furthermore, the scientific resources dedicated to state-recognized medicine and the standards to which it is held put it at an advantage over those medical cultures that have not enjoyed the same scrutiny, in terms of both professionalization and regulation. This has been a complaint from both healers and AIDS activists, who note that the efficacy of antiretrovirals has been tested whereas the effectiveness of traditional medicines remains anecdotal.[59] Attention to the ways in which various communities understand health and illness could lead to more effective treatments as medical practitioners apply culturally appropriate treatments. Culturally bilingual practitioners who can translate biomedical ideas—such as HIV/AIDS prevention—into the appropriate cultural idiom may prove most successful.[60]

SCHOLARLY APPROACHES TO AFRICAN HEALTH AND HEALING

African therapeutics has fallen largely under the scholarly domain of anthropologists, who historically have been much more interested in culture than have historians, who came to social and cultural history relatively late. Some of the first anthropologists in Africa synthesized the ethnographic-like writings of early European travelers and missionaries with their own observations of African societies. Later, anthropologists influenced first by structural-functionalism and more recently by the field of medical anthropology created monographs on singular cultural-linguistic groups—like the Kikuyu—or nation states—like Kenya—based on their own fieldwork. When seeking to contextualize their work in time, however, they tended to paint in rather broad ahistorical strokes. Those scholars who have examined medicine in Africa from a largely historical perspective have tended to focus primarily on biomedicine, inspecting its practice, influence, and utility for and during periods of white rule. Because the study of African therapeutics remained largely the province of anthropologists rather than historians, few works in the field have offered historical depth. There are a few notable exceptions.[61] As a result there tends to be a wealth of synchronic evidence that offers up descriptions of initiation rights, accessories, the practices of African healers, and how such practices fit into the local cosmology. The appearance of information in isolated chronological pockets rather than examinations of change over time, however, necessitates studies that will determine the mechanisms and engines that drive transformation. This brief section looks at some of the major questions driving anthropological and historical research on African health and healing and colonial biomedicine and suggests ways in which this book both builds on and departs from previous scholarship.

Anthropologists of the 1930s and 1940s were very much interested in understanding why, after some fifty years of colonial rule and influence, many Africans still believed in witchcraft. While the assumptions of these anthropologists were highly Eurocentric—they believed that African culture would naturally give way to a more "rational" European one—their conclusions proved quite radical for the time. British anthropologists like Evans-Pritchard, Max Marwick, and Alfred Radcliffe-Brown argued that Africans were not irrational but used a different idiom in which to express their fortunes and misfortunes. They pointed out that witchcraft as well as other forms of health and healing related to ancestral interventions served very practical and functional purposes within African communities. Termed "functionalists," such anthropologists demonstrated how these same phenomena sometimes created better social and community relations, instilled morality, and acted as economic levelers.[62] Later structural-functionalists, such as Absolom Vilakazi (1962), Avel-Ivar Berglund (1976), and Harriet Ngubane (1977), sought to understand the internal and symbolic logic of witchcraft and African therapeutics. These authors placed particular emphasis on decoding the rituals and ritual props associated with African healers in order to understand the internal logic of local cosmologies. Unfortunately, the search for "authenticity" or African "tradition" led many early anthropologists to ignore the role of white rule. When mentioned, white rule served mainly as a backdrop that increased social, political, and economic stresses and consequently the number of witchcraft accusations. Cultural encounters between Europeans and Africans, let alone other groups, rarely served as the focus of such studies. The only early texts that directly addressed the impact of colonialism on African medical practices per se was a journal issue in 1935 that addressed the practicalities of colonial law and the rise in witchcraft accusations.[63] More recent work that has examined witchcraft and witchfinding movements in the colonial era emphasizes that the rise in accusations reflects African responses to colonial pressures rather than a resurgence of African culture.[64] Intercultural encounters are clearly at the forefront of this book, and this text utilizes some of the ideas of structural-functionalist analysis, particularly in seeking to explain the internal logic of chiefly medicines as well as functional aspects of witchcraft accusations, both of which will be discussed further within. It diverges from other works on witchcraft by examining the ways in which this phenomenon challenged not only European sensibilities, but also the rule of law.

In the 1960s and 1970s, medical anthropology emerged as a new field of inquiry that examined how social, cultural, economic, and political factors influenced the health and well-being of individuals within society. Such anthropologists looked at the influence that these factors had on the ways in which people experienced and perceived illness. Working in Zaire, John Janzen (1978) asked how families and individuals made decisions to attend medical practitioners in a multitherapeutic community and why some illnesses seemed to affect only certain communities and not others. In Africa, studies in medical pluralism and culture-bound syndromes abounded during the 1970s and after, and like the earlier structural-functionalists they shared the idea that African healers succeeded based on their mediation of social conflict. Even today medical anthropologists still use the word *illness* to denote a nonbiomedical or "folk" construct, while the term *disease* denotes a biomedical construct.[65] This dichotomy drawn between biomedicine and other medical therapies implies that nonbiomedical diagnoses and therapies reflect cultural values and mores, while biomedicine remains largely divorced from culture, based instead on science. This is despite the fact that, like those of biomedicine, African diagnoses and treatments are based on careful observation and testing of remedies over time. The biological efficacy of traditional African healers' remedies or practices, however, has largely been ignored by anthropologists even while gaining the attention of pharmacologists and venture capitalists. The result is a continued false separation between biomedical and nonbiomedical knowledge, much like the binary drawn by the judges during Ngcobo's trial. Fortunately, more recent medical anthropologists and some historians of science have begun to challenge this dichotomy and show how "western" scientific research and interpretations also reflect the cultural background of the investigator.[66] Feierman points out that both are "forms of ethno medicine," which "are embedded within a system of social relations."[67] While this book does not presume to make judgments on the efficacy of specific African therapies, it does assume that African medical cultures have offered and will continue to offer many efficacious therapies that go beyond just a healing of the social and political body. Likewise the book builds off the work of both medical anthropologists and historians of science in its examination of the ways in which different therapeutic systems—African and biomedical—are culturally constructed.

Historical studies in health and healing in Africa and other colonized areas of the world have gained in popularity during the past decades. Such studies have increased scholars' understandings of the complicated ways in which white rule operated and how it interacted with indigenous peoples. More importantly they have provided a historical context for many of the current healthcare dilemmas in postcolonial states today. Authors have approached such studies from a variety of angles, either by examining specific diseases (tuberculosis or syphilis in South Africa, sleeping sickness in the Congo, and black plague in Senegal), or by investigating certain healthcare providers (nurses and medical aids in South Africa, midwives in the Congo or Sudan, and bio-medical doctors in East Africa). Few, however, have written about healers. While it is rarely possible to group authors into distinct categories as scholars tend to utilize a number of different analytical tools, historical works on health and healing can be roughly divided into political economists of health on the one hand and constructivist historians influenced by the philosophies of Michel Foucault and Antonio Gramsci on the other.

In a seminal article Feierman urged historians to look beyond medical practitioners—African or otherwise—and their influence on African health and investigate instead the broader impact European colonialism had had on the health of the colonized.[68] Political economists of health thus examined how newly introduced epidemics and epizootics, the imposition of colonial legislation, and urban planning and industrialization affected and altered the health of Africans during the era of white rule. In addition to these larger structural factors, political economists have also focused on the availability of biomedical care to indigenous peoples and whether they have access to such care. Randall Packard, for instance, examines how the South African mining industry conveniently sent men home to the rural areas after they contracted tuberculosis in the mines and became unable to work. Not only did these rural areas lack the health infrastructure with which to diagnose and treat this ailment, but this move on the part of mine owners effectively spread this infectious disease to the countryside. Only when the pool of healthy, efficient labor shrank significantly did the owners invest in research and treatment of the disease both at the mines and in the countryside.[69] Although this is an important aspect of the colonial experience, it does not address the other ways in which medicine—biomedical or otherwise—shaped relations between the colonizer and the colonized.

Constructivist historians influenced by Foucault consider science and medicine another site for understanding power, particularly as it operated in the creation and imposition of political hegemony. This included the use of medicine as a blunt instrument of power as well as the power of medical discourse to influence cultural ideas of normality and abnormality, what is healthy versus unhealthy, sane or insane. When applied to the colonial context, historians focused specifically on how colonial biomedicine acted as a "tool of imperialism" that enabled the subjugation of African people and "legitimated" different forms of colonial legislation such as segregation.[70] Maynard Swanson, for instance, looks at how the emergence of bubonic plague in Cape Town in the early 1900s resulted in a medical discourse that linked filth and contamination to African populations. Despite a lack of empirical evidence to prove such connections, politicians used public health measures to serve segregationists' ends—to keep African populations out of the city. Megan Vaughan and Diana Wylie go even further in their investigations and examine the ways in which modern biomedicine constructed the "nature" of Africans within the colonial discourse, both culturally and physically. Such racial and cultural constructions sought to blame African political resistance to colonialism in East Africa on innate psychological disorders and African malnutrition in southern Africa on an African cultural obsession with cattle. Either way, biomedicine sought convenient answers that blamed Africans rather than imply a failure of colonial schemes or take responsibility for creating the condition of ill health.

Given the role that colonial medicine played in the implementation and maintenance of white rule, scholars asked why Africans would elect to use the very biomedical doctors and facilities that both alienated them and contributed to their oppression. By the 1940s many Africans (particularly those in the urban areas) had in fact begun to use biomedical services with greater frequency. Historians influenced by the ideas of Gramsci sought to understand how Africans consented to and collaborated in, but also influenced this aspect of the colonial project. Their attention turned to the role of African biomedical intermediaries—nurses, medical assistants, midwives, and doctors—who helped Africans to make sense of Western medical practices and practitioners. Whereas this scholarship deems colonial biomedicine an instrument of empire, it also sees it as a site of negotiation between African and European players. Shula Marks, for instance, emphasized how middle-class black Christian nurses not only coped with the frustrations of working with white racist doctors, nurses, and patients, but also were largely successful in convincing Africans of the efficacy of western medicine. Rose Hunt asks whether such adoptions reflect African belief in the efficacy of biomedicine as defined by the colonists, or whether perhaps Africans adopted biomedical ideas and interventions because they fit into their own cultural perceptions, local logic, and ideas of authority and prestige.[71]

While anthropologists have focused on African therapeutics, political economists of health and medical historians influenced by Foucault and Gramsci tend to tell the story of medicine and colonialism almost exclusively from a biomedical perspective. Few works have examined the impact of colonialism on indigenous forms of healing or the effect that indigenous therapeutics had on the practice and professionalization of biomedicine in the colonial context. This work uses the insights of anthropologists, political economists, and historians influenced by Foucault and Gramsci, but shifts the focus toward the historical interaction of multiple medical cultures and particularly the negotiation of traditional medicine by patients and healers. By re-aiming our sights on African therapeutics and the way in which healers and patients interacted with biomedicine and Indian therapeutics, my work reveals not only the impact that colonialism had on local forms of healing but also the impact that local African therapies had on the practice and professionalization of colonial biomedicine. Consequently, I problematize what some scholars assume is biomedicine's colonial hegemony and add to the growing number of Gramsci-influenced scholars who emphasize the importance of negotiation between the colonizer and the colonized in the creation of hegemonic discourses and practices.

Though this study is grounded in the history of the Zulu kingdom and Natal, it has broad applicability to the histories of medicine and colonialism. This book does what few medical histories do; it looks at the history of an oral African medical culture over time.

(RE)CONSTRUCTING HISTORIES OF HEALTH AND HEALING

This book covers historical events and processes over a 120-year period that includes the early period of the Zulu kingdom (1820–79), the establishment of Natal, the colonizing of the former Zulu kingdom, increased urbanization, and the professionalization of colonial medicine. It was a period in which the noose of white rule tightened ever more snugly, restricting and radically changing African political, economic, and cultural life. The historical sources utilized for this book thus vary widely depending on the period, place, and subject. Such sources include European travel logs and memoirs; medical journals and books; the notes of medical councils and societies; anthropological studies and notes; government commissions, reports, and indexes; newspaper clippings; correspondence to and from the government; court records; the work of early African writers; oral traditions and testimonies of Africans recorded in the early twentieth century; and more recent interviews with African and Indian healers. Each source presents its own challenges and considerations, though all reflect, some more consciously than others, the time period and conditions in which they were produced as well as the individual concerns of the author or authors. Healers' lack of visibility in archival records and the ways in which Africans and Europeans described healers reflects their marginal legal position. Furthermore, cultural outsiders often used healers as a trope of African superstition, which they blamed for all troubles, from low rates of Christian conversion to the inability of the state to secure a stable African workforce. Below, I briefly discuss my approach toward reading and utilizing these sources as well as specific considerations for reconstructing histories of health and healing.

To determine the usability of evidence one must consider its nature, intended audience, subject matter, and the bias or agenda of the author. Reconstructing history for the Zulu kingdom is somewhat more difficult than for the colonial and later periods for which there are diverse sources. Almost all "precolonial" sources used in part I of this text were written or filtered by whites, and there are far fewer sources available for cross-referencing. Many of these earliest sources tell similar tales about the Zulu kingdom, reflecting not only the close interaction of settlers, but also the period practice of occasionally copying another author's work verbatim.[72] Port Natal settlers Henry Fynn and Nathaniel Isaacs, two of the most noted and cited accounts of the Zulu kingdom, had very distinct agendas. Fynn and Isaacs worked to open trade with the Zulu kingdom—particularly for elephant ivory and hippopotamus tusks—yet on losing their trading advantage and wearing thin the goodwill of their Zulu hosts, they began to advocate Britain's annexation of Port Natal. Their writings thus sought to portray ruthless, tyrannical Zulu kings bent on destroying their neighbors, black and white. Isaacs wrote Fynn in 1832, suggesting how they should write about Tshaka and Dingane: "Make them out as blood-thirsty as you can and endeavor to give an estimation of the number of people that they have murdered during their reign, and describe the frivolous crimes people lose their lives for. Introduce as many anecdotes relative to Chaka as you can; it all tends to swell up the work and makes it interesting."[73] Fynn, like many authors of the period, wrote his memoirs years after his experience in the Zulu kingdom and had his text possibly revised further by an editor's hand.[74] Given such circumstances, a reader may wonder how it is possible to glean reliable information on a topic as potentially inflammatory as witchcraft and the craft of African medical doctors often described in the colonial literature as "witchdoctors."

While obvious paternalistic and condescending sentiments are easily found in this early literature, there are moments and subjects for which precolonial authors were much more reflective and careful in their writings. White traders, missionaries, and travelers tended almost uniformly to view healers involved in sussing out alleged witches with revulsion and disdain, whereas African herbalists, surgeons, and bone-setters were viewed with curiosity and some respect and written about more objectively. This does not entirely prevent us, however, from learning important information regarding witch-finders. For example, when cross-referenced with other sources, Fynn's insistence that those accused of witchcraft met an *immediate* and violent death appears as an exaggeration meant to vilify Zulu kings and chiefs. But despite his bias and erroneous claim that Africans blamed all illness and death on ancestors or witchcraft, Fynn demonstrates a nuanced understanding of

African belief in witchcraft and healing. At the 1852 Natal Native Commission he testified to a number of different types of healers, their gender, methods, and attire, and, in some cases, remedies.[75] Likewise, missionaries who wrote disparaging and exaggerated characterizations of witch-finders for a popular audience wrote other more realistic texts that attempted to analyze and understand the appeal of African healers and the hold that witchcraft had on African communities.[76]

Many of these early writers offered rich details on health and healing as they sought to relate what they perceived as important historical events and details as well as cultural curiosities. Scholars Alexander Butchart and Osaak Olumwullah have recently argued that writers in the eighteenth and nineteenth centuries often adopted the language and approach of natural scientists in their depictions of Africa.[77] Their taxonomical gaze fell not only on the landscape but also on African bodies, and many became keen observers of African culture. Primary sources from this period are often punctuated with elaborate details of African looks, dress, beliefs, and practices related to African medicine. While most sources offer only fleeting observations of health and healing, a few, such as Francis Fynn,[78] Rev. A. T. Bryant,[79] and Rev. Callaway,[80] all of whom were fluent in Zulu, took a special interest in healers and wrote on the subject at length. Bryant's work *Zulu Medicine and Medicine-Men* perhaps best illustrates this genre as it was written as an "ethnological study of the Zulu people from the medical stand point" and published in the *Annals of the Natal Government Museum* in 1909. Bryant's information comes from his own early research on the Langeni people collected in the 1880s as well as the work of J. Medley Wood, curator of the Natal Herbarium in Durban from 1882 to 1913 Such "scientific" interests do not make this work less problematic, and Bryant clearly shows a certain disdain for African healers. He does, however, provide more information to cross-reference, as well as a rather detailed listing of medical applications and remedies.

Cross-referencing of sources is an obvious and essential component for determining the veracity of information about the past.[81] While other texts confirm Fynn's claims that alleged witches were sometimes killed in brutal ways—including anal impalement—during Tshaka and Dingane's reign, they also show that execution was not necessarily immediate and that decisions could be mediated by other healers or changed by circumstances.[82] Successful cross-referencing of course depends on the availability and quality of sources. To get a better picture and understanding of African medical practices and practitioners of this period I have included some sources outside the immediate area and period under study. Given that there are a number of regional similarities in African therapeutics, some of which will be discussed in chapter 1, a case can be made for utilizing sources outside the kingdom, particularly from neighboring Natal. Since most Natal residents were Zulu-speakers, some from the kingdom itself and others who were never incorporated, Natal's Africans shared many linguistic and cultural similarities with those in the Zulu kingdom. I thus utilize some sources that reference the experience of Zulu-speakers from neighboring Natal during the period of the kingdom to verify and elaborate the existence of medical practices found in the Zulu kingdom itself. Similarly, although the nineteenth century was a period of rapid change, certain continuities in medical practices and beliefs persisted. When I find confirmation of such links, I utilize sources from the later period not only to demonstrate the maintenance of some of these practices over time but to flesh out some of the details that are not always evident from the earlier sources. I do not mean to imply by using such sources that all regional medical cultures are completely similar or that those of Zulu-speakers remained static. These cultures were complex and dynamic and regional variations are evident. Certain cultural resemblances and continuities, however, have clearly persisted over space and time. By corroborating neighboring and later sources with earlier primary sources about the Zulu kingdom, I can elaborate and add detail to some healing practices during the earlier period. Some of these later sources include oral sources of the kingdom recorded at the turn of the twentieth century.

Oral sources are an important means of accessing African perspectives on both the early Zulu kingdom and later cross-cultural interactions. These sources vary between oral traditions that tell origin stories of both humanity and the Zulu nation, oral histories that recall the interviewee's personal experiences, and records of narrators' immediate concerns, observations, and desires. Like written histories, one must consider the various influences on oral histories, considering the time in which they were recorded as well as issues of memory, bias, representation, and possible editing done by a narrator or recorder. Many African oral histories for the Zulu kingdom, for instance, were collected by and passed through white mediators. Not only did this mediation affect what information was imparted and how Africans imparted it, but the final recording ultimately reflected what the recorder and editor found of interest and worthwhile. To determine the nature and reliability of an oral text, it helps to consider the relationship between the narrator and the person or persons who collected and produced the final text. For instance, I utilize the collections of James Stuart, a colonial official in the 1890s and early 1900s who recorded many oral histories and also happened to be very interested in the history of the Zulu kingdom. His position and his fluency in Zulu tended to elicit both oral traditions of the kingdom as well as a number of complaints, such as older men's frustration that the colonial government had forbidden the practice of rainmakers despite a persistent drought. Finally, much of the Stuart Archives, originally written by hand in Zulu and English, has now been published by Wright and Webb. This final production involved not only translations to English and explanatory footnotes, but selection by the editors of material they deemed historically useful. Clearly all of this influences the information the reader can glean from such sources. Furthermore, the narrative traditions describing the rise of the Zulu nation are especially diverse and cannot be treated unproblematically. This variety reflects the individual narrator's place of origin, family, religious affiliation, and the time in which the interview was recorded. The James Stuart Archives include nearly two hundred interviews from the turn of the twentieth century, most with African narrators—persons who made up the original core of the Zulu kingdom, those defeated and absorbed into it, and those who fled to the neighboring British colony of Natal. Groups that were politically marginalized by the Zulu kingdom tended to tell stories that emphasized the unnecessary cruelty of Tshaka, while core groups imparted more heroic narratives.[83]

In contrast to reconstructing histories of the earlier period, the later history of European and African medical encounters can be traced more easily given the number and breadth of available sources. Some basic problems, however, arise from the criminalization and limited licensing of healers. Many healers and patients were reluctant to reveal information on the marginally legal and potentially taboo subject of health and healing. Likewise, many African and Indian healers were reluctant to obtain the required government licenses. Inyanga licenses were expensive and difficult to obtain, and Indians found their applications denied on the grounds that they were not Africans. Consequently, many inyangas actively sought to avoid the attention of authorities, and Africans and Indians fearing healers' retributive powers were reluctant to report them. As participants in an illicit activity, unlicensed healers left comparatively few archival records. The main exceptions are the very thick provincial and national files of applicants desiring legal recognition. Other than court records and the letters of a few elite healers, most of our archival and published information on healers for the twentieth century was generated by complaints of white pharmacists and biomedical doctors. Newspapers and memoirs sometimes described healers as cultural curiosities, and, as in the earlier period, certain observant colonial officials and missionaries left detailed anthropological information on healers and their practices. In many cases government commissions, blue books, court records, and memoirs, like many of these other sources, tend to come as synchronic snapshots. Cross-referencing and stitching together these various sources reveals a larger story that enables us to trace both continuities and change over time and space.

Regardless of the period under study, all of these materials, despite and sometimes because of their biases, can be mined for different types of evidence. They can illuminate historical details; popular stereotypes and metaphors; or the concerns, fears, and desires of people living at the time they were recorded. Such evidence is expressed in either explicit or implicit terms. For instance, there seem to be few hidden agendas on the part of Africans or whites when describing the more "mechanical" aspects of Zulu public health; the same cannot be said for the much more controversial issues of witchcraft, "war doctors," medical competition, and the use of human body parts in medicine. Yet stories of medicine and otherworldly power, told by both Europeans and Africans, can be read on a largely symbolic or metaphorical level. Sometimes Europeans are also rather explicit about their own concerns and intentions with regard to African healers, for instance, why the Natal government decided to criminalize healers. At other times incidental or implicit information is given, such as the sex of healers or the fact that healers formed their own military regiments at the end of the Zulu empire. This information is generally more reliable, given that it is not crucial to the argument at hand and may even contradict earlier information given by the same narrator. Clearly, familiarity with the various texts, cultural practices, and cultural tropes of groups, as well as the advent of historical events enables the reader to determine the nature of the evidence. Even with close attention to counterevidence and the ability to cross-reference many texts, there are certain gaps that cannot be filled from the written record.

To investigate some of these lacunae and discover how healers heal today, I conducted interviews with some forty healers from rural areas and small towns in KwaZulu-Natal and the city of Durban in 1998 and 2002. Though these interviews largely supplement and enhance the arguments in this work, they did help to fill in some of historical details missing from archival and published sources. For example, the historical interaction of Indian and African healers was one that was much better fleshed out through the use of oral interviews than through written documents. My interviewees represented a cross section of healers that included many different types of healers (inyangas, *sangomas*, *umthandazis*), both men and women, rural and urban, and African and Indian. The selection was not scientific but was done through various networks of healers, healing associations, and contacts of other researchers. The one continuity was that most healers belonged to a healing association of one sort or another and were fluent in Zulu. A good number of my interviews were conducted in English; however, over a quarter were conducted in Zulu with the assistance of a translator who helped to ease my rough understanding of spoken Zulu. Given that healers have rather specialized and secret knowledge which understandably they may not wish to share with cultural outsiders, I informed each interviewee in the beginning that I was not interested in this type of insider knowledge but sought a wider understanding of how healers practiced within their communities and how they learned their knowledge. Interviews began with healers' descriptions of how they came to their calling, their families' background in healing, how they had learned their healing and medicinal skills, what types of illnesses they could treat, the general types of medicines and tools they used, and how they obtained such medicines and possibly from whom. In some of the interviews I showed historical pictures, mentioned names from the period under study, and showed various patented medicines as a means of elucidating responses.

Though the majority of the interviews were historical in nature, they, like all oral histories, were also inextricably tied to present-day concerns and reflected my own interests and my interviewees' perception of me as a white North American interested in traditional medicine. For instance there was a perceivable difference in some of my follow-up interviews in 2002 as opposed to the original conversations from 1998. The same persons who had been open and informal in our initial interviews proved much more formal and less forthcoming in subsequent meetings. The reasons for this are complex: a changed research environment that saw increased numbers of international researchers and students, a new national promotion of "indigenous knowledge systems," and the professionalization of local traditional healing associations, as well as some of my own changes. In a few instances, "individuals" I had grown to know gave way to healing-association "spokespersons" replete with talking points.

Given these connections between the past and present, let me say a quick word regarding the circumstances in which the bulk of these interviews took place, so the reader can gain some insight into healers' concerns at the time. In 1998 South Africa was experiencing massive urbanization; the HIV/AIDS crisis was beginning to be acknowledged publicly by government leaders, and the stigma of the disease remained quite high; the Natal Parks Board had begun to encourage healers to consider growing muthi gardens; and the prospect of legalizing traditional healers loomed in the near future. Such events and concerns provided a backdrop to the interviews, and consequently some healers implicitly and explicitly expressed concerns that included: fear of losing their "tradition" and hence the need to record their history, a desire to gain academic recognition for their craft, worry over the seeming proliferation of witches and witchcraft, awareness of the environmental impact of overharvesting of medicinal plants, and consciousness of the possible consequences involved in the legalization of healers. Despite all of these more recent phenomena, when I asked healers how their own practices had changed, many claimed to practice just as their forebears had. Specific questions about types of medicines and how such medicines were obtained, however, showed that this was unlikely. I approached oral sources much like other types of sources: I looked for corroboration from both archival and published sources, as well as from evidence offered by interviewees themselves, such as photos, letterhead, and other cultural artifacts.

PART I

Negotiating Tradition in the
Zulu Kingdom, 1820–79

1 ⇌ Healing the Body

Disease, Knowledge, and Medical Practices in the Zulu Kingdom

WHILE TRAVELING THROUGH the most northern coastal territories of the newly established Zulu kingdom in 1822, Henry Francis Fynn fell ill and "delirious" with fever, the dread of locals and travelers alike. Laid up in a hut and awaiting his ship, Fynn recalls being taken to and treated by a male healer and his two female attendants: "On coming into an open space, they lifted me up and placed me in a pit they had dug and in which they had been making a large fire; grass and weeds had been placed therein to prevent my feet from being burnt. They put me in a standing position, then filled the pit with earth up to my neck. The women held a mat round my head. In this position they might have kept me for about half an hour. They then carried me back to the hut and gave me native medicine."[1] In his published diary, Fynn tells his reader that his recovery three days later resulted from this treatment and the African healer who doctored him. One of the first white traders and settlers in the Zulu kingdom, Fynn settled in Port Natal in 1824 and became fluent in Zulu. He not only received medical care from African healers, but claims to have studied their craft and administered African and European therapeutics to Africans and whites alike, something easily confirmed in the historical record.[2]

Because Fynn's diary is the only source that mentions the fire-pit cure for fever, it is impossible to confirm the truth of his story. Used with caution, however, Fynn's published recollections offer a rare glimpse at African therapeutic practice and knowledge during the period of the Zulu kingdom. But they also indicate the impossibility of separating out the African-European encounters that began before and would become more frequent during the "precolonial" period of the Zulu nation (1820–79). This interconnection is important to point out in a chapter that purports to talk specifically about healing practices in the Zulu kingdom. The historical interactions between Africans of the Zulu kingdom and white traders, missionaries, and Natal administrators and settlers reveal both the nature of these intercultural encounters and some of the medical practices and knowledge of persons within the Zulu kingdom. Though the Zulu kingdom remained politically independent until its defeat by the British in 1879, it was not unadulterated by white influences. British traders came to the Zulu kingdom in 1824 seeking trading relations and were granted permission to settle along the kingdom's periphery in the area that became Port Natal (renamed Durban in 1835). Acting as a client chiefdom of the Zulu kingdom, Port Natal provided the Zulu ruler access to weapons and coveted trade goods, but it also attracted Africans who sought to *konza*, or submit themselves to, these new white chiefs (discussed in chapter 3). Initially this proved unproblematic; King Tshaka, the first Zulu king (1816–28), wielded enough power to demand subservience and the return of defectors from his kingdom. Another white presence, albeit a small one with minimal internal influence, was that of missionaries either invited or given permission to practice within the kingdom.

In 1838 a third white community—the Boers—settled along the western edge of the Zulu kingdom in the Drakensberg foothills after defeating a Zulu battalion at Ncome River. A year later they moved closer to the heart of the Zulu kingdom as King Mpande granted land in return for Boer military assistance to overthrow his brother King Dingane. Likewise Port Natal grew, attracting both whites and Africans alike, and eventually became recognized as the British colony of Natal in 1843. Zulu influence over their white neighbors declined greatly, and what had once been peripheral communities now had a much greater impact on events inside the kingdom. These white communities absorbed those wishing to avoid military service, seek political refuge, or earn money outside the grasp of the patriarchs. Even before the British annexation of Natal in the early 1840s, several chiefdoms had left the Zulu kingdom in favor of life under this new neighbor, though this sometimes worked in the other direction as Africans in the second half of the century sought to escape labor and taxation requirements in Natal. Over the years, not only did the presence of Natal provide an escape to individuals and chiefdoms of the Zulu kingdom, but white missionaries and traders from Natal also introduced new ideas and goods into the Zulu kingdom. Whites thus had an effect on the Zulu kingdom both militarily and by virtue of their bordering presence, yet their impact on African therapeutics and the social and political standing of healers within the kingdom was limited. It was only during Cetshwayo's reign (1872–79), with its increased diplomatic relations with Natal, that concessions were made on this topic. Nevertheless these white sources, as well as African testimonies from the later half of the nineteenth century, have helped shape what we know of healers for this early period.

By carefully weighing and using the observations of cultural outsiders such as Fynn and other white settlers in this area in combination with African oral histories of the late nineteenth and early twentieth centuries, this chapter seeks to reconstruct the health ecology of the Zulu nation and the ways that communities and healers within the kingdom conceived of and sought to heal the body and illness. While the primary focus of this section is on the kingdom, that is not to say that there was or is anything intrinsically "Zulu" about the medicines and therapies practiced therein. Rather such knowledge and practices reflect the regional context from which they emerged. The nineteenth century was a dynamic period in which the Zulu kingdom expanded and incorporated various peoples and was affected by its growing imperial neighbor, the Natal Colony. The consolidation of the various chiefdoms into the Zulu kingdom and the introduction of new diseases and epizootics during this period resulted in the sharing of medical ideas and *materia medica* and demanded new health responses. The decline of the Zulu kingdom and the rise of urbanization, migrant labor, and consumer culture resulted in the disappearance of specialized knowledge of herbs, gathering techniques, and medical practices. Specialized and individual treatments of earlier times gave way to general remedies for a general public. Unfortunately, it is not always feasible to pinpoint when such changes occurred; where possible I have highlighted specific dates, though certain generalities remain just that. Despite these constraints I have tried to move away from an old but often cited historiography that has portrayed a static and synchronic notion of cultural beliefs and health practices of this area and often asserted such similarities through the mid-twentieth century. Likewise I seek to correct the false assertion that persons of the kingdom attributed all illness to ancestors and witches. By highlighting cultural approaches to the body and its ailments in the Zulu kingdom, I seek to establish a basis (albeit shifting) from which to observe further transformations and reconstructions of healing "traditions." Likewise, such an exercise helps explain why certain biomedical drugs and procedures, such as inoculation, later came to be adopted or sought after while others, such as amputation or even pills, were rejected.

THE ZULU KINGDOM IN REGIONAL CONTEXT

Familiarity with local flora and fauna gleaned through observation and experimentation and then passed (sometimes sold) from generation to generation among kin, neighbors, friends and cohorts provides a basis for some of Southern Africa's common folk remedies as well as the therapies of medical specialists. Unlike healers today, who may have knowledge of various ailments and remedies, medical knowledge of the past was highly specialized. This applied to individual healers and families, as well as communities in this region of southern Africa. Bryant claimed that this specialized knowledge can be traced back to the end of the eighteenth century, when single herbal remedies were known and owned by a family.[3] In the nineteenth century several colonists mention purchasing healers' secret remedies and that few healers or families seemed to have access to more than one or two remedies.[4] If healers could buy and sell remedies, why did they not own more than one or two? Presumably, selling one's remedies within a smaller community could eliminate one's potential revenue source or pose a commercial threat; perhaps selling remedies to transient Europeans, whose evidence we have of this practice, seemed less problematic.[5]

African medical beliefs and local therapeutics currently practiced in KwaZulu-Natal show remarkable similarities with wider regional beliefs and practices, evincing a long history of interaction and some common origins. Though medical practices and *materia medica* may vary among different cultural groups, southern Africa's local medical cultures share many key attributes. These include similar herbal remedies, surgical and non-invasive therapeutic techniques, and an occupational division between healers who use only herbs and those who heal through clairvoyant means. The area's cultures also historically shared a maxim of no-cure, no-pay, a practice that has largely disappeared in the face of colonial and postcolonial changes. These regional similarities emerged due to three main factors: (1) older commonalities shared and spread through the much earlier movement of Bantu-speakers to southeastern Africa; (2) the consolidation and expansion of various African polities in southern Africa during the late eighteenth and early nineteenth centuries; and (3) the urbanization, industrialization, and labor migration that accompanied colonization and white rule in South Africa. The movement of peoples—whether by force or voluntarily—brought the prospect of new interactions and introduced many new herbal remedies, healing techniques, and tools, as well as apparitions, diseases, and psychological afflictions to southern Africa.

Linguistic and archaeological evidence shows that Bantu-speakers in southern Africa originated in central and east Africa, where they shared a common culture. As Bantu-speakers slowly migrated southward more than two thousand years ago and dispersed, their culture and language diversified and changed, resulting in a large variety of separate chiefdoms. Likewise each group developed its own medical and ritual specialists, learned to use plants that grew locally, and made therapeutic innovations. Evidence of an ancient common medical heritage, however, can be found in several medical word cognates such as *ti* (medicine), *nganga* (doctor), and *ngoma* (diviner). Anthropologist John Janzen convincingly argues that cultures throughout central and southern Africa share "ngoma," a unique historical healing institution that demonstrates linguistic, behavioral, and structural similarities.[6] Likewise, the medical knowledge of these various Bantu-speaking groups expanded and diversified as they interacted with the local Khoisan inhabitants of southern Africa and through the long-distance trading networks that developed from Delagoa Bay through the Basotho Mountains to the Fish River and northward to the Limpopo Valley.[7] Tracking the exchanges and developments that occurred in African medical and healing systems over the past fifteen hundred years, however, is a difficult and arduous task beyond the scope of this book.[8]

On a wider regional level, different groups gained recognition for their possession of unique materials and skills. The Tonga people, for example, manufactured brass,[9] whereas the Cube produced iron hoes and *assegais* (spears).[10] Specialization also applied to medical goods and skills and contributed to regional trade. While some herbal plants grew throughout southern Africa, such as *iloqi (datura stramonium)*, which was and is used to relieve asthma and bronchitis,[11] other plants, such as *isibhaha (warburgia slutaris)*, popular for curing colds, malaria, toothaches, and other ailments, currently grow only in the most northern part of KwaZulu-Natal.[12] The early predecessors of the Zulus, the Ntungwa, were known for *indungulu (siphonochilus aethiopicus)*, "a medicine for chewing or giving [to] children when having a fever."[13] The Zulu chiefdom reputedly introduced *ikhathazo (alepidea amatymbica)* for colds, the Mpondo produced *umondi (cinnamomum zeylanicum)* and genet cat skins, and the Kumalo and Kuze were sought for their *igwayi* (tobacco).[14] Likewise, medical and ritual specialists often came from specific groups. Local chiefs were said to frequently appeal to the Zolo, Tshangala, and Swazi for rainmaking,[15] whereas the Nzuza were sought for their skills as umsutus.[16]

Oral evidence alludes to the presence of several large and smaller chiefdoms between the Phongolo and Mzimkhulu rivers in the late eighteenth and early nineteenth centuries. Many of the similarities in therapeutics throughout southeastern Africa arose during this time as powerful chieftaincies competed, consolidated, and dispersed in the face of the emerging Zulu kingdom (1820) and Boer and Portuguese slave-raiding.[17] While regional trade had ensured a degree of exchange in medical knowledges in earlier centuries, centralized rule brought together and incorporated peoples and healers from around the region in an unprecedented manner. The boundaries of the Zulu kingdom expanded and contracted several times during its existence, but the majority of the kingdom was situated between the Phongolo and Tugela rivers. Other chiefdoms, such as the Mpondo and Tonga, paid tribute to the Zulu kingdom at different periods and shared linguistic and cultural similarities as well as differences.[18] Other African chiefdoms seeking to avoid warfare or subjugation during this period known as the *mfecane* brought local herbal and medical knowledge and practices into other areas of southeastern Africa as they fled north and east.

ECOLOGY AND EPIDEMIOLOGY IN THE ZULU KINGDOM

Comparatively speaking, the Zulu kingdom enjoyed a relatively healthy ecology, yet changes in neighboring Natal brought in new epidemics and epizootics that affected the health and well-being of the kingdom. Situated alongside the Indian Ocean, the land that became the Zulu kingdom encompasses a wide variety of ecological areas: woodlands, lowveld, tropical forests, great rivers and estuaries, rolling hills, and the steep inclines of mountains. Consequently the area exhibits a wide variety in temperatures and rainfall, as well as a high degree of botanical diversity.[19] The land enabled people to grow five varieties of maize, seven kinds of sorghum, fifty-five vegetables, and twenty-five different wild varieties of spinach.[20] Early travelers recorded African cultivation of indigenous and protein-rich millet, sorghum, jugo beans, cow peas, dates, and nuts, as well as greens, round potatoes, pumpkins, melons, berries, and fruit.[21] Vitamin C could be gotten from marula fruit, which contained two to four times the amount of vitamin C of orange juice, but was more likely obtained from the popular beverage *utywala*, a homebrewed millet and maize beer.[22] This potentially nutrient-rich diet could be further supplemented by fermented milk and occasional meat from cattle, goats, sheep, and game. Much of the Zulu kingdom also proved good for stock-grazing, and substantial cattle herds had caused environmental erosion due to overgrazing by the early nineteenth century.[23] Despite the occasional locust invasion and wide ecological diversity, the Zulu kingdom and neighboring Natal led early settlers such as Isaacs to advertise its fecundity: "The climate of Natal is congenial to vegetable life, as is proved by the rapid germination of the seed, after it is sown; the seasons are also exceedingly encouraging to the growth of all descriptions of vegetable productions; the dew, during the intervals of the periodical rains, being extremely fertilising and nutritive."[24] Likewise, W. H. Bleek emphasized how the variety of temperatures enabled settlers to grow tropical fruits at lower elevations while tending Mediterranean crops at higher ones.[25] Such claims were somewhat overblown and meant to recruit white settlement, yet the land clearly proved suitable given that white settlers not only took up commercial farming in colonial Natal, but later acquired large areas of fertile lands in the former Zulu kingdom.[26] While not all areas were equally fertile, such ecological variety allowed for a large number of herbal remedies.

The rough seas that bordered southeastern Africa's coastline, as well as the swath of Lebombo Mountains rising to the northwest, largely discouraged the presence of outsiders such as the Swahili merchants or the Portuguese who visited and settled in Delagoa Bay only a short distance away. The nearby Portuguese presence since the early 1500s contributed greatly to the diversity of available crops in the area long before the British arrived at Port Natal. This included staples like *idumbe/idumbi (cobcasia antiquorum)* tubers, which originated in Southeast Asia, and maize from the Americas. Although this last crop offered less nutrition and drought resistance than local millet and sorghum, it suffered less destruction from birds and yielded large crops up to twice a year.[27] European settlement on the borders of this area, initially in the east and then in the south, also brought new crops of pumpkin, beans, and high-yield sweet potatoes and arrowroot, all of which were eagerly incorporated into the diet of the local populous. Likewise, a number of exotic species of trees and plants, syringe (*melia azadarach*) trees from Asia and tobacco from the Americas to name but two, found their way to southeastern Africa before European settlement in the area. As demonstrated by the Ngcobo court case, a number of exotic species adapted naturally over time to South Africa and grew uncultivated, eventually becoming incorporated into the local diet and pharmacopoeia.

The territory of the Zulu kingdom, much like the Cape and Natal, had a relatively mild disease environment. While some major tropical diseases were indigenous to the area, southeastern Africa's varying climate and topography, as well as human intervention, made this area a much less hazardous place to human health than places like West Africa, which had been dubbed in Europe as the "white man's grave." The relative healthiness of South Africa was, in fact, one reason for its early settlement by Europeans. Given our limited historical sources as well as biomedicine's imperfect understanding of tropical diseases in the early nineteenth century, it is difficult to determine exactly what types of diseases occurred and to what extent within the Zulu kingdom. Early European travelers, who traveled along the coast, tended to group local diseases generically under the term "fevers," which may have been malaria, yellow fever, or typhoid fever. Fynn's diary refers to the "marshy country" along the coast and the malaria that affected persons there as well as the occasional outbreak afflicting those inland.[28] Robert Plant, who wrote about the unhealthiness of St. Lucia, a freshwater estuary, later died of "fever" in this same area in the late 1850s.[29] Similarly, H. Scheuder writes about a feverish epidemic in 1851 that was rampant along the coastal regions of the kingdom.[30] Other common health complaints recorded in the Zulu kingdom included intestinal and parasitic worms, dysentery, and rheumatism.[31] Likewise yellow and dengue fever and bilharzias were indigenous to the area, and gonorrhea, while not indigenous, had been introduced via the Indian Ocean at an early date.[32] Overall, by virtue of the local ecology, human intervention, and military prowess, the kingdom's population enjoyed good health and grew steadily, particularly during the peaceful reign of King Mpande (1840–72).[33]

The coming of European colonialism generally resulted in a deterioration of health of colonized peoples. The years 1880 to 1920 proved the most deadly period for the African continent. Not only did Europeans introduce new epidemics and epizootics, but colonialism helped to spread new and old diseases and created the conditions for malnutrition and disease through enforced long-distance migration, increased labor demands, and the destruction of local forms of public health.[34] In southern Africa the impact of these forces had come a bit earlier. Trade routes throughout Africa and Asia that preexisted European colonialism by a thousand years, as well as Islamic pilgrimages to Mecca, effectively exposed African populations to numerous diseases and enabled them to avoid the virgin-soil epidemics that so drastically affected peoples in the Americas and Oceania. West and East Africa, by virtue of their interaction with global trade via the trans-Saharan and Indian Ocean trading networks, had also experienced many of the same diseases as population groups in Europe, the Middle East, and India. Such exposure ensured future populations were conferred a certain degree of immunity against future epidemics. Bantu-speakers of southern Africa not only shared genetic roots with Africans from West and East Africa, but gained continued exposure via internal trade routes that connected them to the east coast of Africa. This was particularly evident when one considers the repercussions of first contact between Europeans and the relatively isolated population of the Khoisan in the Western Cape. Unlike Bantu-speakers, the Khoisan lacked this natural immunity and died in large numbers due to new diseases introduced by early European settlers.

Although Bantu-speakers and Europeans shared a degree of epidemiological immunity, European settlers did introduce new and devastating diseases, not only from Europe, but also from Asia and the Americas. Given the similar climatic environments in parts of Asia, Africa, and the Americas, global trade enabled the spread of tropical diseases between these three continents. By the middle of the nineteenth century many new diseases, brought by both European settlement and trade and Indian indentured labor, had been introduced to the area of the Zulu kingdom; these included plague, whooping cough, cholera, polio, syphilis, tuberculosis, pneumonia, and leprosy.[35] In 1846 Scheuder claimed that pneumonia had become widespread in the kingdom.[36] Likewise, in 1832 Smith says that he saw Zulus with smallpox scars and that he had heard there had been an outbreak sixteen years previously (1816) among the Mthetwa, a large chiefdom soon after incorporated into the Zulu kingdom. Smallpox had found its way to the Zulu kingdom via Delagoa Bay and the neighboring Tonga, who for centuries had direct contact with Swahili and Portuguese traders and bore the smallpox marks that delineated the trajectory of infection.[37] A less virulent form of smallpox may have been indigenous to the east coast of Africa, but Europeans had introduced a much more potent and deadly form from Asia.[38] A second smallpox epidemic began in the 1850s, also coming from Delagoa Bay and passing through Tongaland and Swaziland before afflicting the Zulu kingdom in 1863.[39] This epidemic coincided with the beginnings of Tonga migrant labor, which passed through the Zulu and neighboring Swazi kingdoms en route to Natal's sugar plantations.[40] This second smallpox outbreak, however, also brought about the adoption of the European smallpox vaccination, which, according to Gibson, was successfully "carried out by people themselves."[41]

Indigenous epizootics were well under control by the times of the Zulu kingdom, but European settlement introduced cattle diseases that would devastate local herds by the end of the nineteenth century. These would have devastating effects on the well-being of southern Africans, who largely depended on cattle for nutritional as well as social, political, and economic purposes. Cattle provided status to men, served as a basis for bartering, and became an important means of arranging marriages. *Nagana*, or sleeping sickness, indigenous to much of central and parts of southern Africa, was caused by bacteria harbored in large wild game and passed on to cattle and people by the tsetse fly, which lived in long grasses and forest zones. Despite the presence of a tsetse belt that ringed the borders of the kingdom and was present in deep river valleys such as the Mfolozi Valley,[42] sleeping sickness was and is rarely reported in this region. The presence of large Zulu cattle herds during the period of the kingdom attests to the success of local inhabitants in controlling this hazard.[43] Lung sickness, which originated with European settlement in the Cape in 1854, had affected Zulu herds by 1855–56. This prompted King Mpande to restrict the importation of cattle into the kingdom with a threat to kill any which entered.[44] While this disease depleted the bovine population, its eventual recession left the kingdom's herds with some immunity to future attacks.[45] Rinderpest, the most deadly of exotic epizootics, however, took place after the decline of the kingdom, between the years 1889 and 1897.

This ailment affected not only cattle, but sheep, goats, and wild buck.[46] When combined with other ecological strains of this period, it had a devastating impact, which led many Africans to question the underlying causes of these new challenges.

PUBLIC HEALTH

To combat old and new diseases, healers and individuals of the Zulu kingdom maintained a complex system of public health. The term and concept "public health" is generally associated with biomedicine and concerned with vaccinations, quarantining, and regulations to ensure sanitation and control environmental hazards. Intrinsic to good public health are the notion of prevention and the use of expert knowledge to promote and protect the community. African therapeutics include many of these same concerns and practices. To understand the practice of public health in Africa, however, we must understand Africa's varying notions regarding the origins of illness. These include preventing illness and death along lines comparable with biomedicine, but also taking steps to appease the ancestors and prevent witchcraft or the violating of community taboos. Historian Gloria Waite, who writes about precolonial health practices in East Africa, argues that definitions of public health should include "rainmaking, identification of sorcerers, and control of infectious diseases, as well as public sanitation works and health education."[47] Likewise, contributors to *The Quest for Fruition through Ngoma* expand the lens even wider to view healing practices in southern Africa that include healing not only the physical body but the social and political body as well. They argue that African healing practices transcend Western categories of "personal, social, political, economic, and ecological," and thus should not be studied in isolation.[48] As Janzen notes, healers desire "fruition" or the "state of bearing fruit," not only literally in terms of food production, but also in terms of procreation, general health, and well-being.[49] By adopting this wider notion of public health, we can see how public-health practices in southeastern Africa varied regionally and adapted to new conditions, but also came to be consolidated under Zulu rule.

The earliest written records in southeastern Africa show that local public health practices were reflected in African architecture, community planning, and a strict adherence to rituals and avoidance taboos.[50] Structural-functionalist anthropologists argue that cultural taboos developed as a means of maintaining both the social and the physical health of community members. For instance, the idea that forests were dangerous places frequented by *umthakathis*, or witches, discouraged individuals from living or visiting tsetse-fly areas.[51] Many of these public-health practices may at one time have been dictated by doctors or their chiefs but became incorporated to such an extent that they developed into general practice and local knowledge. Healers, local chiefs, and the Zulu king played an important role in maintaining African public health in terms of setting new health edicts, organizing treatment for the sick, reacting to epidemics, rooting out witchcraft, releasing rainfall, and, as we shall see in the next chapter, protecting and advancing the body politic. Sometimes these measures and interventions were successful; at other times ancestors and witches proved intractable.

Community planning and architecture reflected local knowledge of environmental hazards and were organized to minimize risk. Within the Zulu kingdom, for instance, kraals and communities avoided low-lying malarial areas and distanced buildings and cattle enclosures from forests and known tsetse-fly zones.[52] European settlers remarked that Africans built their homes on higher ground, generally on a slope with sleeping and food-storage structures at a higher elevation where they were least exposed to the wind and with a cattle kraal inside a central fence. Heavy tree trunks, thorn bushes, and thickets meant to keep out intruders—animal, human, or otherwise—bordered each kraal, and a single gate was closed nightly.[53] With the exception of the king's or chief's kraal, which was well populated, each homestead stood at some distance from the others and contained multiple dwelling structures for family members.[54] Such practices minimized the chances for contamination and the spread of disease. Architecturally, dwellings made of mud and wattle were built in such a way that they prevented exposure to the elements while also providing good ventilation. They were warm in winter and cool in summer. A hard polished floor made of cow dung helped to ward off insects, while roof props, which varied in size from six to twenty feet tall, were constructed from hardwood impervious to termites.[55] An open fire pit placed at the center of such dwellings had the effect of not only warming its inhabitants but causing bilious smoke. The effect could be so great that Fynn remarked it sometimes left "only a foot of clear air, in which its inhabitants lie low and breathe."[56] While this might not have been particularly good for the lungs, it most likely proved an effective means of warding off disease-carrying insects. Finally, because these structures were not rooted to the earth but built separate from a hard elevated floor, they could be picked up and moved in the case of a threatening fire.[57] Each homestead was further fitted with lightening pegs—sticks smeared with medicine said to prevent lightening from striking the home.[58]

While Africans may not have known about germ theory (1870s) they did have specific ideas about disease origins and the spread of communicable diseases. These ideas were unrelated to ancestors and witchcraft, and though not always discussed explicitly, they certainly prompted certain public health measures. Specific examples can be seen with regard to the transmission of epizootics, such as Mpande's banning of Natal's cattle during the outbreak of lungsickness and later Cetswayo's inoculation of royal herds against lungsickness.[59] A less specific example comes from Fynn, who remarks that the homes of inhabitants suffering from dysentery were distinguished by placing a cow's head upon the entrance way. While Fynn includes this example to show the absurdity of how "conceit can cure," such placement may also have served as a mechanism to help people avoid a highly infectious disease.[60] Likewise, with regard to gonorrhea, one of Stuart's respondents claims the "traditional notion . . . [was] not to have sexual connection if one has a bad sore, i.e. open one, as this will delay in healing or be always breaking out afresh."[61] White settlers observed in the 1870s that Africans knew sleeping sickness in cattle and humans was transmitted via tsetse flies and thus sought to minimize contact.[62] Perhaps this same idea of vectors extended to other insects as well. Records show that Africans burned dwellings that had become roach infested and regularly burned grass from around their kraals to minimize the insect population.[63] Likewise, Barter observed in 1855 on her journey through Zululand with an ill brother that those who knew of her brother's illness set fire to the grass on which they had passed. She explained these actions as a means to "prevent the danger of infection, so terrible do they fear the *Imbo* [feverish epidemic]."[64] Other preventative measures taken at the household level involved strengthening members both ritually and medicinally with *amakubalo* herbs after the death of a kraal member.[65] On a community level, when diseases or epidemics occurred with unrelenting frequency, entire kraals or communities would be moved in hopes of a healthier site.[66] On a national level, precautions were taken to ensure the health of the king, whose well-being reflected that of the nation. This included preventing sneezing and coughing in his presence when he was eating and forbidding the reception of persons, articles, or gifts from a kraal recently afflicted by death. White colonists interpreted this as fear that "some infection may be conveyed to the king,"[67] and certainly these actions at both the household and community level point to local precautions taken to avoid the spread of disease and illness.

Another important aspect of public health, promoted by many biomedical public-health specialists in the twentieth century, included hygiene and sanitation. Later European travelers and settlers often used tropes of hygiene to disparage Africans and remark on their level of "civilization," yet early settlers commented positively on the personal cleanliness and cleaning habits of Zulu-speakers. Concepts of what constitutes proper hygiene and sanitation

are to a certain degree culturally constructed but cultural practices nevertheless have an impact on the health and well-being of a population.[68] Africans
were said to bathe daily in the rivers, marking off areas for males and females
and using various materials for cleansing.[69] Senior men prepared their hair by
plucking it from the sides of the head and sewing a head-ring on top that was
softened with beeswax blackened by charcoal. Likewise married women also
plucked the sides of their head and applied grease and red ochre to stiffen the
remaining hair into a knot on top of their heads, sometimes scenting it with
the *tonquin* bean.[70] Such frequent bathing and applications of ochre, which
possesses antivermin properties,[71] most likely aided in the maintenance of
health and reduced the chances for tick, louse, and fecal-borne diseases.
Bathing was followed by greasing of the skin with fat, which, as Fynn explained, "keeps the skin pliant in opposition to a hot sun and sharp winds."[72]
Europeans remarked that dwellings were "kept scrupulously clean," being
daily swept, inside and out, "where neither dirt nor ash were to be seen."[73] Feces
were defecated into holes and covered up so they could not be seen, while clay
pots were often used for urination and later emptied by youngsters.[74] Taboos
and prevention against witchcraft deterred the rise of outhouses or (like the
Asante of West Africa) the building of indoor pit-hole toilets flushed with boil-
ing water.[75] Local cultural beliefs, however, insured safer drinking water by
requiring human refuse be dispersed away from streams and drinking water
secured upstream from where people bathed.

In addition to actions that a biomedical model might recognize as effective
public health measures, local communities responded to misfortune, epidemics of fever, and crop infestation with ritual interventions. One local
practice, called *mtshopi*, involved older unmarried girls who donned plaited
creepers and went out into the countryside and visited kraals where they *bi-
na'ed*, or sang, lewd songs meant to drive off harm. At the end of the day they
threw away their coverings and washed, symbolically ridding themselves of
evil. The age and popularity of this particular practice is not known, though
sources indicate its occurrence during Mpande's reign up through the time of
Cetswayo. By 1902, only a few places such as Mapumulo practiced mtshopi;
perhaps, as one of Stuart's respondents speculated, its disappearance was the
result of an increased missionary presence in the area.[76] The *phukula* custom
used to honor Nomkubulwana, a Zulu goddess of fertility, was another community ritual performed as a precautionary measure to assure a good harvest
and prevent the eruption of feverish ailments. This practice allegedly started
before the 1830s and also excluded the royal family. In this ritual, which anthropologist Max Gluckman was still observing and writing about in the
1930s, a garden was hoed for Nomkubulwana, and girls bucked traditional

gender roles by wearing boys' attire and herding cattle.[77] Why this particular ceremony continued while the other disappeared is an interesting question. Those who initiated virginity testing ceremonies in KwaZulu-Natal in the past ten years claim inspiration from the ceremonies of Nomkubulwana, in which virgin girls were used and thus supposedly tested,[78] despite a lack of evidence of public virginity testing during the nineteenth century. Today's virginity testing, while much contested, makes claims of legitimacy based on a "traditional" and African response to a new public health crisis, namely the rising rates of HIV and AIDS.[79]

Although many practices and beliefs of the Zulu kingdom may have improved public health, there were others that were potentially hazardous. For instance, unhealthy practices such as protecting young and small animals from predators by keeping them in sleeping dwellings may have helped spread certain tick and louse diseases.[80] As well, the practice of drinking and eating from the same pot and the propensity to provide hospitality to strangers increased chances for the spread of infectious diseases. Ironically one of these practices stemmed from health concerns regarding witchcraft and a desire to show that a host had not poisoned the food. While the diet available to Africans proved potentially nutritionally rich, Scheuder mentions treating several cases of scurvy among royal patients at King Mpande's kraal in 1851. This may have been due to the large amount of meat consumed at the king's kraal in lieu of other foods or the result of an increasing reliance on less nutritious maize and European foods. Diana Wylie, who provides a portrait of nutrition in the Zulu kingdom, shows that certain cultural beliefs prevented newborn babies and women from obtaining a nutritionally rich diet.[81] Malnutrition was a problem that certainly increased in the late nineteenth and twentieth centuries as Africans had less access to and labor to tend arable land. Such desperation may also have led Africans to eat the meat of not only slaughtered cattle but those that had died. First commented on in 1846 and mentioned by a number of other white observers, this could be a particularly dangerous practice if the animals had died of communicable diseases.[82]

Whereas many local public-health initiatives were incorporated into day-to-day practices or manifested as taboos, the Zulu king represented the ultimate public health official. As Ndukwana, one of Stuart's respondents, explains, "All people like the land they lived on belonged to the king. If any man got seriously ill, his illness would be notified to the mnumzana [headman], who would instantly report the fact to the izinduna [chiefs] and they to the king. The king would then most likely give the order to consult diviners so as to discover the nature and cause of his illness. A sick man in Zululand was always an object of great importance."[83] In theory the Zulu king and his local chiefs took responsibility for the well-being of their people and surrounded themselves with a variety of different doctors to assist them in this function. While not all illness was brought to the attention of the king, kraal heads had to report illness to their local chiefs. Depending on the social status of the ill person or number of persons afflicted, a report would be sent to the king.[84] The Zulu proverb *inkosi yinkosi ngabantu*—a king is a king by the people— emphasized the reciprocal relationship between a king and his people.[85] In exchange for the labor and loyalty of his subjects, the king provided for the welfare of his people, and his failure to do so could lead people to konza to another ruler. Zulu-speakers who konza'ed white rulers in neighboring Natal thus could not understand why such responsibilities were not also assumed by their new rulers.[86] Another reason sickness and death sometimes gained attention at the highest levels of the state was the link between illness and witchcraft. Illness represented the possibility of persons who sought to destabilize the chiefdom or nation, and consequently chiefs could get in trouble for not reporting illness.[87] Upon learning of an illness, a chief or the king would sometimes provide his own doctors, presumably the best in the area, or send for doctors or medicines from the surrounding regions.[88] In some cases the king provided his own personal medicines.[89] The state of public health thus also represented the metaphorical health of the nation state.

During periods of crisis, such as droughts, epidemics, locust infestations, or epizootics, the king would summon his best doctors and mobilize a national response. One notable medical phenomenon led state healers to connect a number of unexplained deaths to the wearing of a whitish metal (perhaps tin or silver). By order of either Tshaka or Dingane—the sources seem unclear on this point—this metal was banned and collected from around the nation and buried.[90] This shows the reach and power of the Zulu state in carrying out public-health initiatives. Another example, perhaps more typical, were the bands of soldiers who were marshaled to kill locusts during times of infestation.[91] Likewise periods of drought led the king not only to hire reputed raindoctors for the nation but to mobilize people to look for *inkhonkwanes*—herbs and medicine pegs put on mountaintops by umthakathis seeking to prevent rain and thus cause social disruption.[92] Whereas these examples point to a reactive form of public health, a number of preventative measures and rituals occurred during public festivals such as the yearly Inyatela (First Fruits) and Umkhosi (royal) celebrations. At these celebrations, large groups of people from around the nation came to witness and participate in ceremonies that took place within a short span of each other in December and January. At these festivals, the king, as the pre-eminent healer of the land, accompanied by his doctors and regiments, performed preventative measures aimed at ensuring the well-being of the nation and all who lived in it.[93]

African healers themselves provided public health by fulfilling many different functions in the Zulu kingdom. Politically, as we shall see in chapter 2, healers helped legitimate the power of the king and local chiefs; they prepared ritual muthi to bind the nation and/or clan and could pronounce punishments of banishment or death on any outcast or formidable political rival in the kingdom. Militarily, healers played a strategic role in planning warfare, empowering the army, and disabling opponents with powerful muthi. Socially and medicinally, healers negotiated and communicated with *idlozis* (ancestors), found lost cattle, detected thieves, brought rain, set broken bones, lanced abscesses, provided minor surgery, administered curative and protective muthi, and, importantly, unveiled umthakathis—a role which, as we shall see in chapters 2 and 3, made African healers particularly powerful individuals and rather unpopular with missionaries and government officials during the colonial period.

White observers noted in the nineteenth century that there were a variety of terms for African healers. Some of these are recorded in Zulu-English dictionaries, giving the historian an idea as to how early certain terms existed and when others came into favor. Such dictionaries, however, must be used with caution, as the absence of a word in a dictionary does not mean that it was not in use; it may have been overlooked by the recorder. Combined with other sources, however, dictionaries show that each healing specialization had its own healers who were generally denoted by the term inyanga—meaning doctor or specialist. An "*inyanga yezilonda* (*elonda*: sore) was thus a doctor who specialized in healing sores; an *inyanga yonzimba-mubi* (*umzimba-mubi*: bad body)—an abscess-doctor."[94] The naming of healers by their particular specialization seems to show a continuation of Bryant's single-remedy doctors from the late eighteenth century up through the 1870s.[95] Yet we see that the more general term *inyanga yokwelapho* (*elapha*: to treat medicinally) was recorded by Dohne as early as 1857 and defined as a master of administrating herbs.[96] Another term for a general herbalist was *inyanga yemithi* (*mithi*: plural for medicine), a healer who treated bodily ailments medicinally or surgically. Ancestors sometimes helped direct healers to find new remedies, but this occupation was largely learned by apprenticeship with family members. Other types of inyangas that worked at the national level included an *inyanga yokumisa izwe*, or doctor for "making the land stand firm," who treated the national *inkatha* (symbolic grass coil) which secured the strength of the nation.[97] An *inyanga yezinsizwe* (*zwe*: nation) or *umsutu* strengthened the army and nation through medicine. A rain doctor helped not only to bring or deter rain, but also prevented lightning strikes. Diviners, known for their ability to negotiate with idlozis and point out umthakathis, helped to relieve afflictions caused by such agents. In the mid- and late nineteenth century, the more general terms *isanuse* and *isangoma* (the term varied in different localities) were used to refer to any type of diviner.[98]

Depending upon their specialty and area in which they practiced diviners could be referred to by a number of different appellations. An *inyanga yokubula* (*bula*: to beat) seemed the most popular form of diviner. Such healers discovered the nature of illness or misfortune by asking people in a community to beat sticks or assegais on the ground, saying *siyavuma*, "we agree," or *yezwa*, "we hear," in response to the healer's questions. Depending on the forcefulness of the pounding or verbal response of the clients, the diviner was able to learn the relevant details of the case, and if a participant did not smite the ground properly she/he was likely to be reprimanded by the diviner.[99] After determining the nature of the case, the diviner explained the origins of an illness or misfortune, concluding with detailed instructions for its resolution, or sometimes with a referral to an inyanga specialist. Other types of diviners included those who used *impepo*, an herb used medicinally as well as ritually for initiating contact between a healer and the ancestors; such healers appear less often in the historical record. Likewise they laid out the origins of the problem, explained how it had occurred, and offered a solution. These diviners made their reputations by requiring no input from their clients and led Callaway's respondent to lament in the 1870s: "Among diviners of the present time there is no longer any clear evidence that they are diviners; and we now say, they have not eaten impepo."[100] Another less common type of diviner but one which received greater mention than *impepo* diviners were the *umlozikazanas* (whistling diviners). These persons claimed no special medical knowledge but owned *imilozi* (voices of ancestral spirits). Like the impepo diviners, they often required no input from their patrons. Instead clients heard a high-pitched or whistling voice external to the healer that spoke to them and, in cases of witchcraft, allegedly presented material evidence that fell from the rafters in the roof. One of Stuart's respondents relates how his father had attended such a healer in the early 1870s and been told to see an inyanga, who was to prepare and administer medicines in a specific way.[101] Some imilozi, however, borrowed the practice of the inyanga yokubula and demanded verbal audience participation and pounding of sticks.[102] This particular form of divination seems to have been practiced from the time of Tshaka through

Dingane's reign but to have shifted during that time from being practiced on a national level to being practiced in the household.[103] "Throwing the bones," a common form of divination used today by many Zulu-speaking healers, appears to have been adopted at a later period from outside the Zulu kingdom, as there are no mentions of this practice during the kingdom. Instead, this practice seems to have been much influenced by Sotho healers.[104]

When a person became ill, the first line of defense involved not necessarily healers but folk remedies. Only after these proved ineffective would a patient, her friends, or family members, who were generally involved in health decisions, send for an appropriate inyanga.[105] If an illness was indeterminate, they sent for a diviner. A diviner's powers were such that she or he did not always have to observe the sick person in order to offer a diagnosis. In fact a good healer attested to the strength of his or her connection with the idlozis by diagnosing a patient without asking questions. Natal administrators and structural-functionalist anthropologists argued that healers, with the help of their assistants, often had an intimate knowledge of the social, political, and economic nuances of the community. Such insight, they presumed, enabled healers to pinpoint the social cause of illness as they drew on their awareness of family and community tensions and hearsay. Likewise, healers were also careful observers of a patient's symptoms. Depending on the diagnosis, an isangoma would then refer the patient to a healer who specialized in that ailment or treat it himself.[106] If the case involved witchcraft, however, this was reported to the chief or king before any further proceedings. Consultations were said to last half a day or longer, depending on the seriousness of the case. During the period of the Zulu kingdom, a healer did not travel to sell medicines or manufacture medicines in bulk as she would in the twentieth century but collected her muthi fresh from local sources. She also stayed with the patient to oversee the administration of her cure.[107] Whatever type of healer was seen, the client paid an ugxa, or small fee, which in the mid-nineteenth century amounted to a string of beads or a small brass ring.[108] A patient, however, did not pay in full until he or she had healed and fully recovered.[109] Presumably this prevented ineffective healers from gaining prominence and saved patients' resources. Furthermore in the face-to-face communities of the Zulu kingdom, healers knew when their patients had recovered. If a healer's remuneration was slow in coming or particularly stingy, healers were known to exact revenge, which often hastened payment. Healers' knowledge of herbs and their alleged ability to counter the effects of witchcraft was based on presumed knowledge of poisons and spells. For this reason healers were generally paid promptly and were viewed with some trepidation—an uneasiness that still lingers today.[110]

Given the diverse types of healers and various names by which they were called, it is interesting to note that by the late nineteenth century African healers came to be categorized, by whites and Zulu-speakers alike, into two main groups: those who possessed clairvoyant powers—allowing them to communicate directly with ancestors—and those who relied mainly on the use of muthi. The terms *isanuse, isangoma,* and *inyanga* became more general, reflecting changes in the healing profession. Healers of the Zulu nation had begun to expand their knowledge and practice as they encountered healers and remedies from other areas of the kingdom, an expansion that would increase dramatically in neighboring Natal under the effects of urbanization and migrant labor. A major distinction between these types of healers in the early and mid-nineteenth century had been that only isangomas and isanuses could diagnose an illness and that inyangas secured patients through isangomas or through a self- or family diagnosis. Today much overlap occurs between these two groups, though it is generally the isangomas (the term *isanuse* having since fallen out of fashion) who divine and the inyangas who dispense herbs. An isangoma, however, may use muthi for ritual purposes and an inyanga may communicate with ancestors through dreams or diagnose by "throwing the bones." In this sense, once-distinct divisions have become much blurred.

CONCEPTS OF THE BODY AND ILLNESS

Within the Zulu kingdom cultural ideas of the body and illness were similar to those of biomedicine in certain ways, but there were also distinct differences between the two. The heart, for instance, was said to be the seat of the "soul" and to move up and down depending on whether one felt hopeful or fearful. Pointing high up on the throat, considered the resting place of the heart, was a means of showing one's gratitude.[111] As in sixteenth-century Europe, prohibitions against dissection limited knowledge of the internal organs and functioning of the body.[112] Overcoming this key cultural taboo allowed Europeans to gain knowledge of the internal body. Within the Zulu-speaking communities of southeastern Africa, this social convention continued throughout the nineteenth and twentieth centuries. Their limited knowledge of the body's internal functions[113] may have come from comparisons made through the dissection of animals and dead enemies secretly dissected during times of war.[114] Tshaka was said to have killed two women for dissection purposes, one to examine her heart, and the other to see how a fetus lay in the womb.[115] Given that dissection was so abhorred, stories of dissection probably should be taken not literally, but as a means for certain Zulu-speakers to emphasize the inhumanity of war or the cruelty of Tshaka. The contentious nature of dis-section was certainly made clear when district surgeons in neighboring Natal, required by law to dissect those murdered or killed in faction fighting, faced angry relatives in the late nineteenth century.[116]

The causes of illness can be divided into two main categories: those that were seen as "natural" and often infectious, and those that resulted from in-terference from umthakathis and ancestors.[117] This division was important as it determined the nature of one's treatment and which doctor was sought. Most natural diseases fell into the *umkhuhlane* class, which encompassed all "fevers"—enteric fever, malaria, influenza, pneumonia, smallpox, and measles—as well as bad coughs and the common cold. The earliest mention of this term is from Dohne's dictionary of 1857, in which he defines it as "a certain weakness or disability in the human body from the effect of cold." Bryant's definition expands this term to include all fevers and natural diseases. And in the 1970s, anthropologist H. Ngubane finds that Zulu-speakers in-cluded ulcers or decayed teeth, as well as diarrhea and hay fever brought about by seasonal changes, into this classification. It is difficult to tell how far back in time these more recent recorded ideas can be applied or whether they may have been introduced or influenced by Africans' exposure to bio-medical ideas.[118]

Other natural ailments included imbalances in the blood and bile. Blood problems were mentioned quite frequently and were said to result in a multi-tude of ailments. Other "natural" ailments included pain from old wounds known as *isilalo* and swellings.[119] The biomedical doctor S. G. Campbell, writing of local African diagnostics in 1878, states, "Tumours and swellings of different kinds are all said to be accumulations of blood. Thus, I have seen a fatty tumour, ascites, sarcomea, hydrocele, abscess, all diagnosed by distin-guished black physicians as accumulations of blood."[120] On the other hand, excess bile was believed to cause stomach and bowel disorders, chest inflam-mations, coughs, headaches, and general debility.[121] Consequently many remedies involved bloodletting and purging the stomach and bowels. Other natural ailments, particularly headaches and abdominal and nervous disor-ders, were sometimes diagnosed as the effects of a tiny black beetle known as *ikambi*. Doctors were known to suck out such beetles and produce them as evidence of a cure to patients and families.[122] Ngubane noted in the 1970s that another class of natural diseases, known as *umfuso*, included genetic diseases, such as epilepsy, asthma, chronic bronchitis, unhealthy skin, and certain types of insanity. Again, it is unclear if this is a new classification or one that went unrecorded in the earlier period. The term *umfuso* does appear in two 1905 dictionaries, but its definition of "resemblance" does little to clarify this question.[123] Perhaps as Ngubane suggests, the umkhuhlane class itself led contemporary Africans to "a readiness to experiment, to try new medicines, or to discard some for better ones. There is also a general belief that the understanding of this type of natural illness is common to most people, including people from outside Africa. For this reason there is readiness to use curing techniques and medicine of a Western type."[124] As is discussed in Chapters 3 and 4, Zulu-speakers readily tried European forms of medicine from an early period and for a variety of different types of ailments, including problems of witchcraft, which clearly fell outside the umkhuhlane category.

When illnesses, with the exception of genetic ones, lasted for an inordinate amount of time or did not respond to the usual remedies, other causes, namely ancestors or witchcraft, were suspected.[125] A person based his or her suspicion of a disease's origin on how it began and whether it affected only the individual, affected only the individual's family, or was widespread. Diseases caused by the *idlozis* (ancestors) were said to come upon a person quite suddenly, often causing the sufferer to experience unconsciousness.[126] Suspicion of witchcraft might emerge, however, as people observed the uniqueness of their predicament. Isaacs describes such a case in the 1830s: "A body of the people in the neighborhood came to us with pensive looks, and complaining in a pitiful strain that sickness had invaded their families. They seemed to think it singular that they alone should be sick while all the people around them were enjoying good health."[127] Such an occurrence led these people to question the nature of their illness and to seek permission to hire an isangoma, who could presumably determine the origins and resolution of their affliction.

THE ANCESTORS

The idlozis, or deceased, were said to be vigilant in observing and protecting the well-being of the living, and thus much care was taken to honor them and include them in important life events. According to one of Stuart's respondents, a healer might reprimand a patient for not honoring his ancestors, explaining illness as a consequence of such neglect: "You do not praise him, you never refer to him, you give him too little."[128] For this reason when a homestead or the capital of the Zulu nation was moved, residents and doctors took heed to move the kraal or national ancestral spirits to the new site.[129] Idlozis retained the same personality in death as in life and showed their pleasure or disgust with the living by ensuring their good health and fortune or by sending illness and hardship. Ancestral disapproval usually resulted from breaches in social and cultural mores or the breaking of taboos—such as failing to secure purification after killing the enemy during periods of war.[130] Such a diagnosis thus resulted in stern warnings and accusations of "being obstinate and contumacious."[131] Interestingly, many ill wives and mothers were accused of suffering from their own indiscretions.[132] This is perhaps not surprising given married women's external lineage to the household, as one can assume that both living and ancestral relatives would scrutinize their behavior rather carefully. Likewise, women undergoing difficulties in childbirth were often blamed for bringing on their own complications and suspected of inappropriate behavior toward their husbands.[133] In this way, women's illness could serve as an opportunity for men, wives, and neighboring healers to critique and attempt to control women who deviated from acceptable gender norms. The exception to ancestral berating was the twasa illness suffered by those undergoing initiation as diviners. This illness, suffered by both male and female initiates, proved especially useful for married women, who could escape certain social pressures as they returned to their ancestral home for training under a master healer. Furthermore, when such women did return to their husband's kraal their social status had greatly increased, and "deviant" behavior was more likely to be excused.[134]

The general recommendation for appeasing the idlozis involved the slaughtering of a beast that was then specially treated with muthi. Because slaughtering was such a common request for appeasing the ancestors, bouts of family illness were often described by saying that the ancestors were hungry.[135] At other times, however, idlozis might insist that rituals improperly performed in the past be re-enacted. Given that African health generally deteriorated in the late nineteenth century, as a result of both new and old ailments and defeat by the British and the resulting civil war, Stuart's respondents complained at the turn of the century that ancestors could be exceedingly demanding in their requests for sacrifices. When demands exceeded people's resources, families strategized. They did not automatically sacrifice a beast but made a necklace of its tail hair for the sufferer. If the sufferer then improved, the beast was slaughtered; if not another diviner was sought.[136] Stuart's respondents reflected a growing frustration with ancestors by the end of the century: "Even if there is only one beast in the kraal, the *idlozi* (through the *isangoma*) would direct it to be slaughtered . . . Because of the obvious stupidity of sacrificing the welfare of the living for the sake of the dead, the *amadlozi* came to be called *izituta* [fools]."[137] It is interesting to note in the Stuart Archives that ancestors rather than healers or Europeans bore the brunt of blame for increased sickness and misfortune. While this may reflect some hesitancy to blame colonial policies before an official, Stuart's respondents argued that the idlozis had turned their backs on the people.[138] In frustration Africans in Natal and the former Zulu kingdom began in the late nineteenth and early twentieth centuries to investigate alternative sources of well-being, namely Christianity and biomedicine.

UMTHAKATHIS AND WITCHCRAFT

When sacrifices for the idlozis failed to bring about the expected recovery of the patient, people began to suspect poisoning or witchcraft. Accusations of witchcraft, however, depended on the type of illness, its severity, and the social and political circumstances that accompanied it. Umthakathis were people who wished to inflict harm on others, usually due to jealousy or anger.[139] This might include envy over the number of cattle or goats a neighbor owned, the plentiful crop of a clanswoman, the attention a co-wife received from the husband, or anger at a person's failure to pay back a loan. Armed with the knowledge of powerful poisons and the possession of helpful familiars, umthakathis were said to have the ability to harm and afflict death on community members. Those who desired to harm others but lacked the power and knowledge of this art allegedly could hire such a person to assist them. Umthakathis could afflict harm by mixing *insila,* or the body dirt of an intended victim, with specific muthi that was usually delivered by a witch's helper. In most cases the umthakathis need not even be present given they were aided by familiars, such as the *impakha* (cat), *umhlangwe* (a small brown nonpoisonous snake), *umkhovu* (mute dwarf), *isimfene* (baboon), and *isikhova* (owl).[140] Reverend Owen paraphrased King Dingane's (1828–40) description of such individuals: "They went out in the dead of the night carrying a cat under their arms; that when they got to the house of a person whom they intended to bewitch, they sent this cat, who was a sort of little messenger, into the house; that the cat brought out either a bit of hair, or a bit of the cloak, or something else belonging to the unhappy victim, which the witch deposited in some secret place under the floor of her house, and that in consequence the object of her malice in due time became sick."[141] One description of witchcraft included the use of insila and certain herbs. Buried by the fire, this mixture would cause death by preventing a person from urinating, causing the bladder to "burst open and fill the stomach."[142] Another form of witchcraft included laying *umbulelo* (a poisonous concoction) along paths or in a kraal targeting specific persons who would then fall ill.[143] Food or water could be poisoned with powerful herbs, or, as one of Stuart's respondents alarmingly claimed, an umthakathis could inflict death on a victim merely by chewing medicine, spitting it out, and stabbing the air with a doctored assegai while naming the intended victim. Umthakathis were also accused of withholding the rain to damage the entire community. At the end of the Zulu kingdom, the same respondent claimed that umthakathis had begun to practice the Sotho skill of killing individuals with lightning.[144] Given that the Sotho were renowned rain doctors, it is not surprising that such skills, albeit negative, might have been attributed to them.

In an effort to guard against the alleged and real effects of umthakathis, a number of customs and taboos emerged. These included requirements such as partaking in a meal or beverage before offering it to one's guests, scattering and hiding one's feces, and guarding the grave of the newly deceased. The Qwabe clan, which was closely related to but not a part of the Zulu kingdom, allegedly attempted to control use of powerful medicines by placing restrictions on who could and could not become a doctor.[145] Other restrictions on the use of muthi (common folk medicines aside) prevented most persons from haphazard experimentation. Ordinary people mixed muthi only for their ancestors and did so in the presence of family members. Mixing herbs alone, explained one Stuart respondent, was said to cause the idlozis to harm or kill one's relations.[146] These taboos not only helped safeguard the practice of African healers but could be seen to protect persons from witchcraft, as such behavior brought the immediate notice of one's neighbors. As Fynn argued, such behavior meant "on sickness or death prevailing in any locality, a person whose actions had previously raised suspicions which had spread throughout the neighborhood, is now suspected of being the guilty cause of such a calamity."[147] Indeed such beliefs might then be asserted and confirmed before a witch-finder—an isangoma, isanuse, or inyanga yokubula.

The belief that a witch did not have to be present in order to pose a threat but could do so from far away or with the help of an emissary meant that many people could find themselves accused of witchcraft. It was said that the only means for curing an illness caused by witchcraft was to discover who had inflicted it and the means by which it was done. Accusations thus often included the discovery and destruction of a witch's charm. Given the fear of witchcraft, witches discovered by diviners were often killed or expelled from the community. As Fynn states, "Criminals of this class were looked upon as enemies of the State, so as in war, not only such persons but all related or connected to them in any way must be put to death."[148] While this does not technically hold true, its sentiment surely reflects the perceived danger of such individuals. Witchcraft remained a problem for communities throughout the period of the kingdom and increased during the ecological and political disasters to come.

Other "non-natural" illnesses not associated with the idlozis or umthakathis included the *isibethelo*, or love charm, often used on young women.[149] It was believed that a love charm would induce a woman to madness if she refused the man who administered it.[150] One concoction recorded by Bryant at the turn of the century consisted of cuttlefish, *umanyanya (psychotria capensis)*, *umginakile (Asclepias meliodora)*, *ulukuningomile* (not listed), and *uzililo (Stapelia giantea)* plants, leopard's fat, and the spittle of both the woman and the man who desired her. When placed beneath a projecting rock on a precipice, this isibethelo was said to ensure a woman's affection.[151] Women also used love charms on men, though most often to secure attention from their husbands. Christina Sibiya claimed in the 1920s that it was common for a father to send his daughter to an inyanga before her marriage. There she acquired "love philters, secret potions, and devices by which she can hold her husband's love."[152] Persons using love charms outside marriage undermined the hierarchy of the community and were often sentenced to death, though in 1852 Fynn claimed that the use of love philters sometimes brought about only a fine.[153] Marriage was a family affair involving male elders, and although a woman was given some choice in her marriage partner, love charms prevented the family from making a rational choice about a marriage partner and potentially "devalued" the woman through madness or premarital sex. Should a young woman run away with a man, her family faced losing *lobolo*, or "bride wealth." Love charms also enabled young men (and perhaps collaborating girlfriends) to circumvent the accepted norms of society, providing leverage in a society ruled by elder men. Complaints about love-charms increased amongst fathers and chiefs as young men increasingly escaped the grasp of patriarchs by going to work in the urban areas in the late nineteenth and early twentieth centuries.[154]

Other types of "love medicines," seen to uphold the traditional values of the community, however, were condoned. This included the administration of *umsizi*, a poisonous concoction, to a man's wife. This poison did not affect the woman, but an untreated man who had sex with her was said to become violently ill. Given the difficulty in ascertaining who had or had not been treated, the threat of umsizi allegedly discouraged men from having sex with married women. The husband of a woman treated with umsizi, however, remained unharmed so long as he took the prescribed antidote.[155] The resulting and rather painful disorder, known as *jovela*, however, was said to be easily cured by doctors, raising questions about umsizi's effectiveness in thwarting unwanted sexual behavior.[156]

It is difficult to determine the botanical knowledge of Africans during the early nineteenth century as compilations of Zulu medicinal plants were not produced until the end of the century.[157] In 1909 Bryant described 240 medicinal plants used by Zulu-speakers.[158]

Muthi, however, did not include just botanical sources, but, as was indicated in Ngcobo's 1940 trial, could also include minerals, seawater, and parts of animals. Bryant wrote that muthi that produced certain illnesses was believed to cure the same symptoms.[159] Such descriptions read much like homeopathy, and one wonders if knowledge of this alternative western medicine had an impact on white writers' understandings of this African medicine. In the early twentieth century, physician James McCord described how ingredients for muthi were chosen by virtue of properties that mimicked the affliction being treated. For instance, bleeding from the nose or mouth was treated with the "bark of trees which have juice like blood, and parts of an animal which bleeds readily when touched."[160] The opposite principle also applied. Thus a person suffering from nervousness and fear might be treated with the heart and eyes of a lion and the fat and flesh of powerful animals.[161] While certain current muthis may fit this description, there are plenty that do not.

In addition to those available from medical specialists, folk remedies included medicinal plants readily recognized and obtained by family members. Common medical knowledge included treatments for injuries, most umkhuhlane illnesses, and less serious ailments such as indigestion, worms, or headaches. If a person was injured, the wound would be irrigated with cold water, either through the teeth or a reed, and covered with an herbal salve.[162] People treated bone fractures by supporting the broken area with wooden splints or dog bones and securing them with a grass fiber or strip of tendon.[163] Serious broken bones and joint dislocations, however, required the help of a trained healer. The fern, inkomankoma, was used by Africans and Europeans alike to rid people of worms and is mentioned for this use in Fynn's diary.[164] The term inkomankoma seems to refer to a variety of different types of ferns, including both indigenous and exotic ferns, all of which were used for this purpose.[165] Common herbs used for indigestion included umondi (mondia whitei), indawo (cyperus esculentus), and umhlwazi (catha edulis) and were often worn around the neck for one to nibble on as needed.[166] Older women generally assumed the responsibility of caretaker for the sick and consequently knew more about illnesses and their remedies than did other community members.

Specialists were consulted to treat more serious ailments. Most but not all the remedies listed in the last section were employed by inyangas for "natural" causes that did not necessitate the intervention of ancestors. Vapor baths, either moist or dry, provided a frequent means of curing fever, malaria, or respiratory ailments. *Pungula*, or steaming, involved enclosing a patient in a large skin, blanket, or grass mat while he or she crouched over a pot of medicines caused to boil by the insertion of red-hot stones. Afterward the doctor sprinkled the patient with this same water while it was still very hot. Bryant states that this cure was also used to treat people with insanity.[167]

Like European colonists of the period, Zulu-speakers blamed blood imbalances for many illnesses, and, like their European counterparts, African doctors practiced bloodletting or cupping as a means of regaining balance. Bloodletting was used in cases of rheumatism, swellings, and local pains. The doctor created an incision, or *inhlanga*, over the afflicted area, applied an *isilumeko*, or cattle horn, and then used the mouth to create suction and successfully draw out the blood.[168] "Clots of blood" were blamed for head wounds and treated by making an incision on the head to remove the "blood clot."[169] Blood imbalances could also be remedied by numerous blood cleansers made of herbal infusions. Many urban Africans would later abandon these herbal cleaners in favor of patented blood cleansers sold in European chemist shops.

The common practice of injecting medicines was practiced on young and old and involved making a number of small incisions over a painful or swollen spot and then rubbing in various remedies. Fynn remarked at seeing such scarifications on Tshaka and others and mentioned that such procedures were frequently used for shoulder and groin pains.[170] Terms for this practice, known as *gcaba*, *zawula*, or *jova*, came to be used synonymously with the biomedical terms "inoculate" and "vaccinate." Bryant claimed in 1905 that *jova* was a recent word of English origin. The fact that this practice already existed in local African therapeutics and Europeans couched explanations of vaccinations in culturally translatable terms enabled future colonists to impose vaccination requirements fairly effectively among much of the African population in Natal and Zululand. This was in contrast to Natal's white population, which remained skeptical and often noncompliant.[171]

Healers treated more serious broken bones by creating an incision over the broken area and rubbing in parts of the *umtombe* (fig) tree, banana tree, and *umahlogolosi (urginea altissima)*. These were said to cause the bones to unite, which, along with a splint, seemed to be very effective. Even today when Africans go to the hospital for a cast many also stop at the herbal market for the herb *umahlangenisa* (lit., cause to join together) *(urginea delagoenis)* to ensure a positive outcome.[172] Another means of treating broken bones or severe dislocations involved the administration of a plasterlike cast. Gardiner, an early settler, explains the remedy healers used to cure John Cane's disjointed and broken arm:

> Several men having assembled at the place, with a native Esculapius at their head, a deep hole was scooped out, and then partly filled with pliant clay; the whole arm, with the hand open, and the fingers curved inwards, was then inserted, when the remainder of the clay that had been prepared was filled in, and beaten closely down. Several men then steadily raised his body perpendicularly to the incased arm, and drew it out by main force . . . This, I understand, is the usual practice among all these tribes; and is said to be effectual.[173]

This practice, common among Xhosa-speakers, seemed to be practiced only in the far western areas of Natal.[174] It is unknown whether this particular practice was utilized in the Zulu kingdom.

The fear of umthakathis and poisoning led to a liberal use of both preventative and curative emetic therapies. Likewise they were used for ailments of the throat, lungs, and chest.[175] Fynn claimed that Tshaka took them several times a day for prophylactic purposes: "It was not taken from necessity or with the idea of throwing up any poisonous properties remaining in his breast."[176] There is evidence of other kings participating in this same practice, if not daily then at least for important royal occasions. When Tshaka was stabbed, his own doctor administered purgatives (possibly out of concern about a poisoned assegai?) and washed the wound with "decoctions of cooling roots."[177] Purgatives were also given for severe abdominal complaints. Bryant mentions both croton oil and the exotic jalap as common emetics. "A piece of the bark, half as large as one's thumb, is pulverised in half a cupful of milk or broth, and the mixture drunk."[178] In addition to jalap's or *jalambu's* use as an emetic, people came to utilize its tubers for love charm emetics, for smoking fields as protection against lighting, and as an antisyphilitic herb.[179]

Other forms of therapeutics involved the application of herbs through smoke inhalation, infusions, poultices, enemas, suppositories, and manipulation. Headaches or fevers were often treated through the inhalation of smoking roots or snuff.[180] Women were given an infusion known as *isihlambezo* to fortify their pregnancies.[181] Powdered cuttlefish bone was used to remedy weak eyes.[182] Snakebites were cured by *isibiba*, an emetic and powder made from poisonous snakes that was placed over the bite and which Fynn claimed was effective if administered within four hours of a bite.[183] Boils, abscesses, and swellings were treated with an *ithobo*, or poultice, which usually involved the application of a heated bulb of onion or wild garlic.[184] Herbal enemas were also quite commonly used for bowel complaints,[185] and urinary infections were treated with suppositories to the vagina or urethra or with medicine blown through a hollow reed through the penis to the bladder.[186]

With regard to surgery in the Zulu kingdom, Fynn details healers mending ruptured testicles through a combination of surgery and manipulation.[187] Gundersen, a missionary, watched Usihajo (Cetswayo's chief surgeon) operate to remove a lump and then provided sutures to the patient.[188] Head surgery was sometimes done for headaches that resulted from old head wounds. Leslie reported in 1868 that the hair was shaven from the head, the healer "then cut into the bone, scrape[d] well for about five minutes and during the operation [had] . . . water constantly squirted from the mouth."[189] A Stuart informant claimed this practice was commonly done to remove a clot of blood which caused head pain.[190] Perhaps one of the most impressive surgical operations performed was recorded by Lugg in 1947, long after the end of the Zulu kingdom. In this case an injured woman had been scalped, but as the skin was still attached at the back of the head, the inyanga stitched the skin back into place using the mandibles of a large black ant. "As these were applied and closed to hold both pieces of skin together, the ants' backside was snipped off, leaving the 'calipers' firmly in position." The woman was said to have recovered in a month or two.[191] Although this surgery took place in the twentieth century, it is interesting to note what was either African knowledge of suturing or the application of local knowledge to new biomedical ideas.[192]

CONCLUSIONS

Many Westerners and cultural outsiders have held on to the notion that Africans attribute all illness and accidents to "witchcraft" or ancestral disapproval. In examining the knowledge and practices of African patients and healers in the nineteenth-century Zulu kingdom, we can see that this was not the case.[193] Africans have long recognized that common physical ailments and injuries could be treated with medicines alone and without appeal to one's ancestors or accusations of witchcraft. Africans had definite ideas about illness and therapeutic techniques that were rooted in local understandings of the body and bodily functions. On the other hand, Africans may have had good reason to fear the jealous rages of their neighbors and threats of "witchcraft" manifested through apprehensions of poison. Colonial court records from neighboring Natal from 1883 to 1947 indicate that a number of Africans were convicted of administering poison during the late nineteenth and early twentieth centuries, something that will be commented on further in chapter 3.[194] African public health and local medical knowledge and practices of the Zulu kingdom reflected the development of African belief systems as well as observations of local medical successes and failures.

The possibility for exchange of medical knowledge increased in southeastern Africa as chiefs and kings amassed wealth and incorporated new populations under their rule in the nineteenth century. During this period Zulu kings hired the most powerful healers in the area as well as those from the surrounding region. Some of these healers stayed at the royal residence, while others came when summoned or participated in national events such as the Inyatela and Umkhosi ceremonies. Such occasions provided an opportunity for healers to observe each other's remedies and practices. White writers who provided medical service to various Zulu kings often commented on the presence of numerous doctors at important cases. Fynn, who was often called to doctor various royal residents, admits to copying remedies that he learned from watching inyangas at such occasions.[195] While there was an effort to safeguard intellectual property—Fynn remarks that healers often deliberately added unnecessary roots and leaves to conceal their active ingredients[196]—it is also reasonable to assume that a certain degree of medical exchange occurred. This exchange enabled healers to offer a wider variety of remedies to their patients, moving beyond the period when healers offered only one or two special remedies.[197] The centralization of numerous populations and healers under Zulu rule increased the availability of herbs and knowledge of ailments and remedies beyond the circle of the king's renowned healers. Likewise, the establishment of a battalion of isangomas during the reigns of Mpande and Cetswayo also ensured further collaboration.

The rise of more general terms for healers, rather than their identification by specific ailments, reflects this growing knowledge base of healers. The categorization of inyangas and isangomas/isanuses or the conjoining term "traditional healers" adopted in the twentieth century is a manifestation of Africans' colonial experience, which only further intensified encounters with healers from various areas. It is also a reflection of colonial administrators' favoritism of one group of healers over another, a topic that will be discussed in further detail in chapters 2 and 3.

2 ∽ Healing the Body Politic

Muthi, *Healers, and Nation Building in the Zulu Kingdom*

COMMUNITIES WITHIN THE Zulu kingdom maintained a complex system of public health that involved maintaining not only the corporal body but also the body of the nation. Oral histories regarding the role of muthi and healers and the rise of the Zulu kingdom indicate the cross-cultural connections of medicine and power. Healers provided medical, ecological, social, political, and military assistance to the nation. Each king and chief in the kingdom relied on his or her own healer or healers to help obtain and maintain political power. Furthermore, the judicial system relied on healers to determine guilt for petty crimes and witchcraft as well as for resolving the occasional dispute over political succession. The close relationship between muthi and healers and political power sometimes created tensions between healers and rulers. Again, definitively tracing change over time during the period of the Zulu kingdom is difficult. Furthermore, given that the majority of sources for this chapter are oral sources recorded in the late nineteenth and early twentieth centuries, they not only give us a glimpse into this particular time and subject but are more definitive about how Africans interpreted such events at the time they were recorded. This chapter thus utilizes these largely oral histories to gain access to both periods, as they are crucial if we are to try and understand what was happening during this early period and appreciate the reaction of white settlers and the colonial government to African healers and therapeutics.

According to oral traditions of the late nineteenth and early twentieth centuries, the rise of the Zulu nation in the late 1810s resulted from not only the pure cunning of Tshaka Zulu but the acquisition and use of chiefly medicines. These medicines and their exploitation by doctors and chiefs are alleged to have changed the outcome of three core incidents that were essential to the emergence of the Zulu kingdom. In these and other stories we see that medicine acted not only as a tool of empire, in this case a Zulu one, but as an important aspect of social control. Medicine healed the body as well as the body politic. Below I have summarized the story of the rise of the Zulu nation as told by Jantshi ka Nongila to James Stuart in 1903. Jantshi claimed to have learned these stories from his father Nongila, who allegedly had shared with him the Zulu nation's official *izibongo*—the poems and history of the nation composed by the nation's *imbongi*, or praise teller. Nongila, who served as a state spy, presumably heard and witnessed the reciting of the izibongo first hand. Jantsi also claimed his father worked closely with the Zulu chief Senzangakona and the Zulu kings Tshaka, Dingane, and Mpande before he moved to Natal to escape his duties. Clearly such information as told to Stuart was aimed to alert the listener to the presumed authenticity of such tales. While Jantshi's stories may differ from any formal izibongo, they do reflect someone sympathetic to the rise of the kingdom and at least suggest the ways in which people and perhaps those closest to the state sought to explain political change. These narratives are also some of the earliest recordings of these state-building stories and indeed share many similarities to other such accounts told within the Stuart archive. While there may be some unique aspects of this story, it is also largely representative of the many oral narratives of Jantshi's day that demonstrate the importance of medical discourse in the imposition of Zulu political hegemony.

According to Jantshi, Tshaka was the estranged son of Senzangakhona, the ruler of a nominal Zulu chiefdom in southeastern Africa in the early nineteenth century. Tshaka, who proved himself as a warrior, came to serve as the military commander in the neighboring Mthetwa chiefdom and was much beloved by his chief, Dingiswayo. It was the collaborative efforts of this chief and Tshaka that resulted in Senzangakhona's death and Tshaka's seizure of the Zulu chieftaincy. Tshaka had sought vengeance on his father in retaliation for Senzangakhona's chasing him and his mother out of the Zulu chiefdom years earlier. Dingiswayo, who looked upon Tshaka as his own son, agreed to help him and gave Tshaka the muthi necessary to poison Senzangakhona. Dingiswayo then invited Senzangakhona to his kraal to participate in a night of festivities and competitive dancing. At the conclusion of the evening, Senzangakona and his men retired to their sleeping quarters in Dingiswayo's compound. While they slept Tshaka climbed up on the roof and washed himself with the powerful medicines given to him by Dingiswayo. When this muthi leaked through the roof and onto Sezangakona, he awoke and sent his men to determine its source. Learning that his son Tshaka was responsible,

Sezangakona became overcome with fear and fell ill that night. The next day he headed home, succumbing to his affliction a few months later. As a result of Senzangakona's death, Dingiswayo appointed Tshaka commander over the small Zulu chiefdom.[1]

The successful expansion of the Mthetwa and the unrest they caused made many of their neighbors wary, including Zwide, leader of the large neighboring Ndwandwe chiefdom. When the Mthetwa attacked the Ndwandwe, the two armies fought all day until sunset without resolution. In an act of seeming reconciliation Zwide offered the Mthetwa chief his daughter in marriage, and Dingiswayo accepted. Zwide, however, had conspired with this daughter and instructed her to secure a bit of Dingiswayo's semen to put into a snuffbox and return to him. This she did, and Zwide then mixed the semen with strong muthi as a means to overcome Dingiswayo. Having been "doctored," Dingiswayo soon appeared before Zwide in a confused state and unaccompanied by his army. Zwide used the opportunity to seize Dingiswayo and at the urging of his mother instructed that he be killed.[2]

On hearing of the demise of Dingiswayo, the man Tshaka considered not only his chief but his adopted father, Tshaka called forth his finest doctors to avenge the death. The three doctors concocted a plan in which they would approach Zwide, claiming that Tshaka had expelled them and asking for his protection. Having obtained such "refuge" they would then poison Zwide and his army, enabling Tshaka to overpower the Ndwandwe. The plan was adopted, and true to their word they lived with Zwide for three months, during which time they poured medicine into the spring from which Zwide drank and administered muthi where Zwide walked and along the paths as they escaped on their way back to Tshaka. On arriving at Tshaka's kraal, they prophesized that when the moon was full Zwide's army, under the effects of their muthi, would appear before Tshaka with shaven heads. Upon seeing Tshaka's army, however, the Ndwandwe would throw down their shields and fall writhing to the ground. The prophesy allegedly occurred as told, enabling the easy defeat of the Ndwandwe armies. While Chief Zwide managed to escape, the Ndwandwe army was devastated and his chiefdom scattered.[3] It was the defeat of the Ndwandwe chiefdom, as well as a similar victory over the Mthetwa, that enabled the expansion and emergence of the Zulu kingdom.

Fascination with the rise of King Tshaka and the Zulu kingdom led African and white alike to record the oral traditions that detailed its emergence. South African historians and popular writers, however, largely ignored the tales of muthi and the role of doctors as agents of change to explain how Tshaka succeeded over other contending chiefdoms.[4] The oldest historiographical explanations for Tshaka's achievements included those that focus on the force of his personality and his military acumen, his supposed development of the short stabbing spear, the horseshoe configuration of his regiments, and the toughening of his armies through running barefoot and abstaining from sex.[5] Scholars from the late 1960s highlighted the material preconditions of the Zulu kingdom, arguing that state expansion resulted from population explosions that accompanied transformations in agricultural production, from ecological changes, or from struggles to control the ivory trade at Delagoa Bay.[6] More recently scholars have argued that state building resulted from pressures exerted by slave traders from the east at Delagoa Bay and Boer settlers from the west who kidnapped Africans and forced them into slavery.[7] Today's historians agree, however, that many of the technological and strategic innovations attributed to Tshaka in the nineteenth century were widespread throughout southeastern Africa by the late eighteenth century and certainly before Tshaka's rule. Yet Jantshi's stories clearly elide these material, ecological, and slavery themes in favor of histories that highlight political personalities and their manipulation of political opponents through the use of powerful muthi and clever doctors.

The likelihood that these stories occurred exactly as recorded in this or other versions of Zulu oral history is slim. Because healers and chiefs were so revered and played such important political roles, common persons often attributed great powers to them. Furthermore, stories of the Zulu kingdom and its early years are particularly contested. Not only were various groups unwillingly absorbed into the kingdom during its initial expansion, but later civil wars and the kingdom's defeat by the British often led members of subsumed chiefdoms and the newly emergent *kholwa* (Christian) class to explain the rise of the Zulu nation in very different terms. In addition to the Stuart Archives, other interviews with Africans conducted by Callaway, Bryant, and Heinrich Filter reflect some of these same issues, and all reflect the popular culture of the times in which they were recorded.[8] In the case of the written African traditions about this time period, there are the works of Magema Fuze (1922) and Thomas Mofolo (1949), who both came from missionary backgrounds, and more recently Mazisi Kunene (1979), who was a modern-day imbongi of the Zulu kingdom and held a doctorate in literature. Of these texts, Mofolo's is by far the most sensationalized—a story of human greed with a moral lesson nominally based on Tshaka's life. Kunene's and Fuze's texts more closely follow in the vein of other oral narratives sympathetic to the state.[9]

Despite the diversity of existing traditions, stories, and memories, all highlight the crucial role played by healers and muthi in the consolidation of powerful local chieftaincies and the rise of the Zulu kingdom in the late eighteenth and early nineteenth centuries. My point in reciting Jantsi's text and others is not to assess their reliability per se, but rather to use them in combination with other sources to provide a window into the general perception of practices, beliefs, and relationships of healers, medicine, and power during the building and reign of the Zulu nation. Finally, I aim to explore how the content and circulation of these stories provides insight into the imagination of the Zulu nation-state and popular culture of the mid- to late nineteenth and early twentieth centuries.

One thing that these stories clearly reveal is that within popular memory in the late nineteenth and early twentieth centuries, the primary motives for state building in southeastern Africa were personal vendettas and power-seeking. The means employed—poisoning and supernatural manipulation—were culturally acceptable methods for political maneuvering and positioning. Muthi referred to medicine that not only healed or destroyed the physical body but helped chiefs and kings to usurp political power from rivals and kin. Zulu-speakers generally cited any transfer of political power outside the line of succession as evidence of the use of muthi. Muthi was in itself deemed responsible for political success or failure. Acquisition of proper muthi and the powerful doctors who administered it were considered essential to a chief or king's rise to and maintenance of power.

These stories also reflect the cultural beliefs in and about witchcraft as applied to political leaders.[10] As we saw in the previous chapter, at the homestead level unexpected illness, death, or misfortune could be blamed on an umthakathi who acted out of jealousy and/or anger as a means to seek revenge or gain power over rivals. These were persons who either used dangerous muthi to literally poison others or were credited with the ability to cause harm through the assistance of familiars and their knowledge of witchcraft. Likewise, people assumed that political rivalries, which engendered similar feelings and desires and involved political leaders armed with powerful medicines, would result in the death and misfortune of rivals. Zwide's securing of Dingiswayo's semen, for instance, mimics that of an umthakathi who sought power or revenge over others by securing insila or *izidwedwe* (material containing body dirt). Such substances were believed to hold a person's essence, which when mixed with muthi enabled one to manipulate and overpower that person.[11] The actions of chiefs and kings resembled witchcraft, and Ngubane found in the 1970s that chiefs were sometimes called umthakathi as an idiomatic means of referring to their power.[12] Rulers' use of such techniques did not carry the same connotations as witchcraft at the local level, because rulers' actions were presumed to benefit the community or nation. Chiefs and kings were for all practical purposes exempt from prosecution for witchcraft within their respective boundaries. Within the wider Zulu nation, however, chiefs could be and were accused of witchcraft by the Zulu kings' doctors.

Evidence throughout the period of the Zulu kingdom shows that chiefs derived power to rule from chieftainship medicines. Such medicines were not to circulate among the masses but belonged exclusively to the chief and ensured the health, strength, and longevity of his or her reign and by exten-sion the health, strength, and longevity of his or her people. Once a king or chief had acquired power it was necessary for him or her to be ritually in-stalled in this position. When introduced to the community a new king or chief was called to bathe him- or herself in the medicines of the kingship or chieftaincy.[13] Mpande was said to wash with water of *ubulawu*, which Bryant later lists as a love charm in 1905.[14] Such medicines ensured that the king's words sounded sweet to his listeners, much as love charms were meant to guar-antee the receptiveness of a sweetheart.[15] Possession of chieftainship medicines signified legitimacy to rule, and obtaining the strongest of such medicines en-sured power over one's rivals.[16] An allusion to this medicine was made when Tshaka bathed himself in medicines on Senzangakhona's rooftop, a bathing ritual usually performed at a new chief's initiation by his or her most prestigious healers. While hereditary ties provided some confidence in a new leader, the initiation of ritual bathing emphasized the will of the ancestors.[17] This muthi was said to facilitate communication between a chief or king and the ances-tors—a force more knowledgeable, powerful, and far-seeing than mere humans. Thus Tshaka's figurative or metaphorical bathing in medicines given to him by a powerful chief suggests not only Dingiswayo's support but Tshaka's initia-tion as a new chief. In other versions of this story, Tshaka is said to cast a shadow over his father, causing his illness, which is another metaphor used to describe one with medicines more powerful than one's own.[18] Zulu authors Fuze and Kunene are more direct, claiming that Tshaka overcame Senzan-gakhona by obtaining kingship medicines from a powerful doctor.

Chieftainship and later kingship medicines—Tshaka (1816–28) elevated these rituals to the national level—helped chiefs and kings to maintain power through their ability to acquire "the power of foresight, divination, or of prophe-sying future events."[19] King Tshaka was said to have predicted white colonial domination, whereas King Cetshwayo allegedly foresaw the British attack on Zulu forces prior to the Anglo-Zulu War.[20] Callaway's informant describes an *isithundu* as a narrow-mouthed watertight vessel made of grass that when filled with ubulawu gave chiefs and kings the power of the strongest doctors. Through such a vessel, they could allegedly foresee the outcomes of battles and subdue their enemies, both foreign and domestic.[21] In another version of Zwide's defeat of Dingiswayo, a Stuart respondent maintains that Zwide "stirred up a mixture of medicines so that it frothed over. [After which] he saw Dingiswayo in the medicines."[22] Likewise, chiefs, who reputedly saw the success of Tshaka and his armies within the isithundu medicines, took heed and fled with their people.[23] As one of Callaway's informants emphasized in the mid-nineteenth century: "All the knowledge of the chief is in this vessel. If he wishes to kill another chief, he takes something belonging to that chief, and puts it in the vessel, and practices magic on it, that he may kill him when he has no power left."[24] The loss of such an important vessel caused great consternation in the nation; if not found, it was said that "the diviners point out many men, and many are killed."[25] Whether real or metaphorical, its loss represented a loss of political power and the attempt to find it showed a king's attempt to reassert his power through force.

While there are few stories of chiefs or kings healing the body themselves, they nevertheless promoted themselves as the preeminent healers of the land endowed with political, medical, and divine powers.[26] For example we see that Zwide doctors Dingiswayo himself rather than relying on healers as Tshaka does. Because of the direct correlation between medicine and power, the king, chiefs, and healers jealously guarded these medicines. Only the chief or king was allowed to use chieftainship or kingship medicines, and anyone else found using or possessing such medicines was liable to be punished. Muthi provided not only the means, supernatural or otherwise, to challenge political leaders, but the gumption as well. Ndukwana, one of Stuart's respondents, states, "It was also heard said that a man who stirred up medicines cast his 'shadow' on the king."[27] Interpreted literally, Callaway's respondent claimed a chief or king could feel in his body a person who used great medicines. He claimed that they would begin to sweat and feel as if they were bearing the full weight of this person. This may explain why in one version of events Senzangakhona succumbed to illness after falling under Tshaka's shadow. After a strengthening with medicines, Callaway tells his reader, a king could approach such a person and demand to view his muthi. Upon examining and inquiring about unknown medicines—where they were obtained and their proposed usage—he would then place them among his own.[28] Clearly it was the prerogative of the king or chief to own and control all medicines within his jurisdiction.

Of these healers, the umsutus of the nation proved particularly important during the consolidation of powerful regional chieftaincies and the emergence of the Zulu kingdom. This role continued with sporadic military excursions against Zulu neighbors and later in battles with the Boers (1838), for two brief Zulu civil wars in 1856 and 1883–84, and against the British in 1879. These doctors developed war strategies and performed ceremonies with *intelezi*, medicine to strengthen and protect the nation and its regiments from the enemy. They doctored the troops with ritual muthi as was the case in the battle of Isandlwana (1879) when the doctor, according to a Stuart respondent, "made all those with guns hold their barrels downwards on to, but not actually touching, a sherd containing some smoking substance . . . in order that smoke might go up the barrel. This was done so that bullets would go straight, and, on hitting any European, kill him."[29] During the Anglo-Zulu War, British soldiers also reported finding Zulu bullets surrounded by thin gauze and packed with poisonous muthi.[30] Umsutus also healed the wounded and ritually cleansed warriors so that they could re-enter Zulu society at the end of battle. An umsutu could poison the enemy directly, as Tshaka's doctors poisoned the Ndwandwe wells, or could manipulate him from afar by collecting his insila and izidwedwe and mixing it with muthi.

In addition to reading these stories of medicine and power on a largely symbolic or metaphorical level, we can also see that they served several functional purposes as well. By crediting muthi for their political success or military victories, political leaders helped to create a sense of fear and awe among their people. King Mpande, for instance, allegedly referring to the Zulu victory over the Qwabe, stated it was "not by our valour but by our drugs (izintelezi) [protective medicine] alone."[31] Medicine in and of itself was dangerous and had to be kept in the right hands. Political leaders benefited from the belief that they alone possessed the most powerful muthi, as it could be applied not only to their political enemies but to their subjects as well. Paulina Dlamini, a servant to Cetshwayo, told her biographer how King Cetshwayo caught a fugitive of his kingdom by placing an article of the deserter's clothing on the royal inkatha.[32] Likewise, Mpengula Mbanda claimed that chiefs ensured their subjects' submission by sitting on the inkatha, metaphorically ensuring their power over the people.[33] These stories of muthi and power referring directly to Mpande and Cetshwayo's reigns helped maintain social order by encouraging subjects to view rebellion or escape as a futile gesture against the supernatural power of kings and chiefs. Kings and chiefs not only had the power of divination and therefore were all knowing, but possessed the wherewithal to capture deviant persons.

The wide circulation of these stories could also have been a way for Africans to explain rapid political change in which they had little say. In earlier days Africans could vote with their feet by picking up and going to live with another chief. During the period of the late eighteenth and early nineteenth centuries, however, such movements were limited by expanding political chiefdoms. Although some people fled to Natal and regions outside southeastern Africa, many others found themselves unwillingly absorbed into a new nation-state and forced to pay tribute, or had their sons and daughters drafted in the army or state service for years on end. These stories may have provided some solace to those subjected to the expanding Zulu chiefdom; they could blame the use of muthi for their defeat rather than a lack of military fortitude. On the positive side, the muthi of the king or chief could also be used to protect his or her population from outside aggressors and natural disasters. Certainly it was better to align yourself with someone who had access to powerful medicines than not.

Within the new kingdom, as in the old chiefdoms, healers furnished the king not only with muthi, but also with a ritual authority that lent legitimacy to his reign. Chiefs and kings depended on healers to provide chieftaincy or kingship medicine with which to install and inaugurate them, muthi with which to divine, antidotes against sickness or poison, ritual protection from local or national enemies, and impartial judgments of criminal and civil cases within their jurisdiction. In other words, doctors healed both the corporeal and political body of the chief or king. Rulers sought out the most celebrated healers: those who had curatives for hard-to-cure ailments, reputations for bringing rain, and success in warfare. Dingane was said to show great faith in his sangoma, who found hidden things, discovered those who caused illness by witchcraft, and could explain why regiments were late in returning home.[34] Zulu kings recruited healers within and outside the kingdom. Tshaka, for instance, was said to have hired Gasa rain doctors,[35] and Mpande to have housed at least ten doctors from various areas of the kingdom while relying mainly on four to doctor the army and himself.[36] Cetshwayo hired a Basotho healer to doctor his army during the Anglo-Zulu War.[37] Politically and militarily, healers were an asset to those in political power.

Having emerged as the most successful and powerful of the local chiefdoms, the Zulu expanded by successfully integrating diverse clans and communities through political and cultural assimilation. The employment of medicine and healers was instrumental in transforming the Zulu kingdom from one that based its political mandate and operations on lineage affiliation to one that supported a diverse but centralized state. Many of these medicinal practices to secure and preserve political power were not new but incorporated and eclipsed local traditions and practices by elevating them to a national level. African healers would play a central role in many of these new and expanded practices and rituals. One of the most important institutional changes implemented by Tshaka involved the creation of a national *amabutho*, or state army. The amabutho of the nation worked much as it had on the chiefdom level, only one's allegiance was given to the king rather than one's chief. The drafting of young men and women from diverse clans into age-regiments helped not only to foster a new identity but also to provide one of the main mechanisms by which the king established authority over his subjects. By maintaining the amabutho the king had not only direct control over numerous regiments with the power to discipline insubordinate groups but also the power to increase the nation's wealth by ordering the amabutho to raid cattle, to hunt elephants for ivory, or to exact tribute from smaller surrounding chiefdoms. The amabutho, integral to the success of the new nation, were assimilated into many of the nation's rituals in an aim to foster allegiance.[38] Likewise, the king appropriated the Umkhosi, or Festival of First Fruits, during which the chief tasted crop samples and held a ceremony that marked the commencement of the harvest season to a national level. Tshaka declared this prerogative his own and expanded the festival of first fruits to include not only the tasting of first crops but the reviewing and ritual emboldening of his warriors, the announcement of new laws, orders for regiments to marry, and the strengthening of the nation through applications of muthi to the national inkatha.[39]

Healers who always had important roles to play in the chiefdoms became integral to the Zulu kingdom's national ceremonies and rituals. One such ritual that greatly helped to foster new bonds of Zulu nationalism involved the national inkatha, a ceremonial coil created by the king's *indunas* (chiefs), doctored by the king's healers, and woven to about fifteen to eighteen inches in diameter. This was in addition to his own personal inkatha, which was buried with him at death. The national inkatha was sturdy enough to support the weight of the king when he sat on it during the national festival of the first fruits and during wartime. It was said to consist of *umvithi* grass (*eragrostis plana*), cloth, and strengthening muthi such as the plants of *umabophe* (*acridocapus natalitius*), *usangume* (not listed), *umatshwilitshwili* (*plumbago auriculata*), and *imfingo* (*stangeria eriopus*).[40] The use of these herbs signified the connection between healing the body and healing the body politic. Within the African therapeutics of this area, umvithi and umatshwilitshwili were used to heal broken bones, umabophe was said to avert anger when a transgression had been committed and avoid danger, and imfingo allegedly protected warring parties by cleansing and protecting the body from harmful spirits.[41] Furthermore the inkatha included insila of the regimental houses and vomit of the Zulu regiments collected during an *umbengo*, or ritual vomiting ceremony, which generally occurred during the Festival of First Fruits. During wartime, however, an umbengo occurred prior to each military campaign, and the king's doctors helped to secure military victory by tying the ritual vomit mixed with grass into the inkatha.[42] As a sacred object, the inkatha was not to be spoken of and was brought forth only on the occasion of great feasts during which the king, assisted by his doctors, performed ceremonial ablutions while sitting upon it. During wartime, the king was expected to protect his nation by sitting on the inkatha for as long as his forces were engaged in battle. According to Dlamini, Cetshwayo sat on the inkatha during the battle of Isandlwana, but as he learned of his forces' victory, he left it with greater frequency. When a report came back that one unit was unable to defeat a small British garrison at Rorke's Drift, Dlamini, who most likely witnessed or heard such accounts herself, reports that the mothers of the nation "reproached Cetshwayo severely. They put the blame on the king for not having occupied the inkatha uninterruptedly."[43]

In addition to the inkatha's pivotal role in protecting the nation, it was also used as a means of fostering national unity. The king's healers presented the national inkatha and publicly strengthened and "doctored" it at the Umkhosi. During this time, national insila was collected by young men from around the kingdom and bound into the national inkatha with the sinews of a slain ceremonial ox. Dlamini tells us that this insila was said "to embody the power and energy of the nation. These 'soul substances' were found in the 'body dirt' of the populace, but especially that of chiefs and the king himself . . . Small samples of these specimens were then incorporated in the magic coil whereby the soul of the nation, represented by it, became enlarged, strengthened and rejuvenated."[44] The powerful symbolism of the inkatha was lost neither on 1920 Zulu nationalists, who formed a cultural society under this same name, nor on the society's successor, the Inkatha Freedom Party. Dlamini described the symbolism of the nation's inkatha to Reverend Filter in 1925: "Its contents symbolized the unity of the nation and all the values associated with the king's ancestors. The properties and magic prowess of wild animals embodied in this coil were transferred through it to the people. Tshaka greatly strengthened the power of the *inkatha*. He subjected a large number of tribes but formed them into a united people by collecting bits from the *izinkatha* of vanquished tribes and particles from the bodies of slain chiefs and embodying them in his own coil."[45] In referring to the binding process, one of Stuart's respondents stated, "The inkata's purpose is to keep our nation standing firm. The binding round and round symbolizes the binding together of the people so that they should not be scattered."[46] The inkatha not only bound the people together under the protection of the king but also represented the people, who likewise supported the weight of the king. As a popular Zulu proverb emphasized: "Truly a chief is a chief according to the people, not according to the grass that he possesses."[47]

The symbolic nature of the inkatha helped foster a new Zulu identity, but its importance also meant it needed to be jealously protected lest it fall into the wrong hands. When not in use at national ceremonies or during times of war, the inkatha remained with the nation's *inhlendla* (ceremonial barbed assegai) in the *enkhatheni* (literally, place of the inkatha) and was guarded by an old woman. This most sacred place was located within the *isigodhlo*, or royal housing area for the king's wives, concubines, and female bodyguards. This was one of the most regulated and protected spaces of the Zulu kingdom, and any man found entering the isigodhlo allegedly faced the punishment of death. Given that older women often performed ceremonial offerings to the idlozis in the home, it is not surprising that the inkatha, the most powerful symbol of the Zulu nation, was housed and protected by women. Women rarely held positions of political power, except for those related by blood to the king who were appointed as chiefs. Consequently we see that the king not only more readily trusted women to guard the nation's inkatha but relied on female guards to protect himself as well.[48] During the Anglo-Zulu War, the British burnt down the king's palace and destroyed the national inkatha, which had been passed from one Zulu king to the next.

A king's healers not only supposedly ensured subjects' alliance to the nation, but in one case could be appealed to in an effort to resolve chiefly disputes. A story told by Mpatshana and alleged to have occurred during Mpande's reign (1840–72) involved a succession dispute between two sons of chief Manqondo of the Magwaza people. Qetuka, the elder brother, complained to King Mpande that his father favored his younger brother Mpeyana as the chief son rather than himself. In the evidence brought before Mpande, it was determined that Mpeyana was favored by the Magwaza, who described him as both generous and hardworking. Mpande nevertheless requested that Manqondo follow the law of primogeniture by appointing Qetuka as his chief son. Manqondo acceded. Subsequently Mpeyana requested an abandoned kraal site from Mpande so that he could step aside and let his brother assume power. Impressed by Mpeyana's gesture of good will, the king gave him the land and peoples of Nomazocwana. Initially pleased, Qetuka soon soured as his people followed Mpeyana to Nomazocwana. Again Qetuka requested Mpande's intervention. After reprimanding Qetuka for causing his own problems, Mpande nevertheless ordered Mpeyana and his people to return and sent his own doctors to resolve the dispute. The king's healers assembled the peoples of Magwaza and stirred up medicines and sacrificed beasts to promote acceptance of Qetuka's rule. While this story does not indicate whether the medicines used were chieftaincy medicines, it attests to belief in the ritual authority of the king's doctors and in the power of their muthi to at least figuratively unite a fragmented community that had otherwise ignored the orders of its own chief.[49]

Medicine of this magnitude was clearly a powerful weapon in the arsenal of any chief or king, whether it fostered political unity, protected the nation, or physically eliminated his or her opponents. In the case of one chief it is said to have caused both his elimination and his moral disgrace by allegedly leading him to commit incest with his sister, become ill, and eventually be eaten by a leopard.[50] Chiefs and kings took precautions to protect themselves from the muthi of both possible rivals and subjects. They hired healers with powerful muthi, such as Tshaka's umsutus who poisoned the Ndwandwe wells, but depended on healers to provide preemptive antidotes as well, such as the one which Tshaka digested with each meal.[51] But even medicine was not always enough, as is suggested by one Zulu chief who appealed to King Mpande for permission to kill his own son on the speculation that the son was buying muthi to use against him.[52]

THE JUDICIAL SYSTEM

Evidence shows us that healers also played an important role within the judicial system throughout the period of the Zulu kingdom, and it was their involvement within this important social institution that alerted and alarmed colonists in neighboring Natal to the political power of healers. The judicial system of the Zulu kingdom consisted of two main branches—the *ibandla* (king's council of elder statespersons) and the *umhlahlo* ceremony, which was presided over by isangomas. The first process, one of open debate, was fairly transparent to outsiders like the British. In fact many early white traders and settlers of Natal and Zululand adopted these same open judicial procedures with Africans within their settlements. The second process, however, probably seemed quite opaque and arbitrary and was eventually banned by British colonial administrators. The distinction between these two processes became increasingly important with the encroachment of colonial rule and destruction of the Zulu kingdom in 1879.

Both processes generally took place on a local level before proceeding to the king. During a local ibandla, for instance, a chief would call together the *izikhulu*, or principal men and women of the district, to discuss a case. Testifying before the Native Laws and Customs Commission in 1881, Shepstone stated that these trials allowed for careful cross-examination by any person. "Any man may sit with the assembly and cross examine and give his opinion. The accused himself is severely interrogated, and the truth or otherwise of his answers is tested by reference to persons who may have been incidentally mentioned in his statements."[53] According to King Cetshwayo, when conducting a trial the chief "does not hear the evidence and then go home and consider it, but he sits there and listens to the assembly talking about the case. Then they ask his opinion and he says, 'I think so and so.' If they don't agree with him, they give their reasons."[54] Together the chief and council decided on an appropriate punishment. If they differed, then the case was taken to the king, where it was discussed among the chief men and women of the country.[55]

Whereas the ibandla dealt with a variety of types of cases, almost all cases dealing with death and illness were referred to an umhlahlo. Many of the cases that came before the national umhlahlo involved the illness or death of important subjects, often on the suspicion that such events had resulted from foul play or the work of umthakathis. Mandatory rules to report illness and death to the king or chief signified not only the king or chief's concern for the physical well-being of his or her subjects, but concern for the political stability of his or her domain.[56] A sick person, particularly a man or woman of political significance, threatened possible political disorder through political rivalry or the underhanded maneuverings of an umthakathi. In either case this was a direct affront to a ruler's power and necessitated immediate and direct action. The umhlahlo, however, could also be used to hear other cases, such as those of theft, succession disputes, or when two people of high stature failed to find consensus on an important issue.

A national umhlahlo took place after healers had gathered evidence at the local level. At such an umhlahlo the king summoned representatives from each household and community concerned as well as distinguished isangomas. According to Ndukwana, a healer who gave great details from Dingane's reign on, summarized a typical umhlahlo. It generally took four days to summon isangomas, who often traveled long distances from across the nation to try a case. Issues of objectivity and fairness led to a preference for healers unfamiliar with the case at hand. Each doctor came with a group of armed followers to protect him or her from those involved. The umhlahlo began as the disputants and representatives of the community gathered in a circle. Each sangoma then entered the circle separately in an effort to divine the nature of the problem, with the most celebrated sangoma entering last. Ideally this method was intended to keep the isangomas from knowing what the other doctors had divined; in practice, however, this was not always the case. The results of either the ibandla or umhlahlo would then be told to the king, who acted as final arbitrator and determined a settlement or punishment. Those "sniffed out" as umthakathis were considered dangerous and unredeemable and often faced imminent death, particularly in the early years of the kingdom.[57]

Both Henry Fynn and Stuart's respondents mention the holding of umhlahlos, or large trials, where people of different kraals gathered before the isangomas or isanuses and *bula*-ed (participated in a call and response with a healer). We see that this practice continued into the late Zulu kingdom. Beleni, one of Stuart's respondents, describes such an umhlahlo he witnessed in 1871. The case involved the ill health of Chief Madhlodhlongwane ka Jaja of the Dhludhla and took place at Nodwengu.[58] Although King Cetshwayo was not present (chiefs and kings rarely presided over such events), at least two of his army regiments and a number of his isangomas participated. Starting around one o'clock in the afternoon, the ceremony lasted until sunset. The regiments formed a circle, which doctors entered one by one. Such doctors carried a shield and assegai with which they threatened their audience in an effort to provoke vigorous participation and elicit the cause of Madhlodhlongwane's ailment. "One doctor entered at a time. He would strut about, sniffing out with his nose. He would then call on them [the participants] to strike and bula. They would strike the ground with short, heavy sticks, and dust would rise. He (the doctor) has his assegais and is standing. He says, 'Tshayani' [strike] (and they respond) 'Izwa' [listen]. And as soon as he smells out the person, he leaves the circle in a hurry and runs away."[59] The doctors varied in their actions. Some pointed out a person or persons as responsible, while others seemed unable to come to a clear decision; some took a short time, while others took longer. In this particular case, one of the king's isangomas, Mpezulu ka Ntshona of the Magwaza, "contradicted the other doctors who smelled out different people. Mpezulu reputedly said, 'They [isangomas] speak falsely; those of the house are alone harming one another; they are contesting the patrimony [i.e., the right to succession]'"[60] At this point Madhlodhlongwana apparently recovered, vindicating Mpezulu's claim and resolving the father's illness.

During the early period of the Zulu kingdom, most suspected cases of witchcraft and theft were tried by royal or chiefly isangomas at a national or chiefdom umhlahlo. Hypothetically all local cases that had been tried and provided sufficient evidence of guilt were followed by a national hearing. Likewise, in theory no one within the Zulu kingdom could be killed without the permission of the king. As Stuart's respondents point out, chiefs asserted themselves in the period following the first Zulu civil war, killing off followers without the king's permission. As one respondent says, "The izunduna used to kill off commoners and then report to the king that they had 'taken away' this or that person for practicing witchcraft."[61] Chiefs asserted this power during periods of instability in the kingdom and at its collapse.

AMBIGUOUS RELATIONS BETWEEN PEOPLE, HEALERS, AND THEIR RULERS

An important strain in these stories about medicine and power is their abuse at the hands of healers and leaders alike. Although healers were highly regarded, a number of stories indicate that kings and chiefs had lost control over isangomas and that it was these healers "who played havoc with the king's subjects."[62] It is unclear when such concerns began, though the oral traditions suggest that it was rather early. One of the earliest that we know of says that Dingane warned of such dangers and allegedly said of his people, "When I leave them, when I die, they will be destroyed by izinnyanga through this little house of Songiya's [the mother of Mpande]."[63] Consequently, later kings such as Cetshwayo were said to appoint men "to be [their] eyes and ears, for the purpose of warning anyone that was about to be smelt out to run for refuge to [them]." Both Mpande and Cetshwayo allowed banishment rather than death for some of the accused, and Cetshwayo established a place of refuge called *izwe labathakathi* (literally, the land of witches). At this place the accused could live and either prove their innocence or show that they had been rehabilitated by avoiding future allegations.[64]

On the other hand, many of Stuart's respondents alleged collaboration between chiefs and healers who sought to accuse "especially wealthy men of being abathakathi, so that, as these had many cattle, the chiefs would be enriched."[65] Such stories critique rulers who abused their power and indicate that Zulu chiefs had often acted independently of the king and for their own benefit. Stuart's respondents largely emphasized the economic advantages of witchcraft accusations, rather than describing them as a means of political maneuvering, which was generally attributed to muthi. The relationship between the two, however, was not entirely lost on Zulu respondents such as Baleni: "For when a man had repeatedly been given presents of cattle and these had multiplied, people would come and say he had enormous herds and accuse him of overshadowing the king. Upon this he would be accused of being an umthakati and then be put to death and his stock seized."[66] This stock would then be added to the king's or chief's herds. Lunguza, who had many relatives killed by King Dingane, accused the chiefs of deliberately making witchcraft accusations in an effort to increase the size of their own herds. He claimed that chiefs would act and then tell the king that a person had been killed on account of witchcraft, though apparently chiefs were careful not to practice such tactics on the great warriors of the nation.[67] In practice, witchcraft accusations acted as a mechanism that not only enriched chiefs but ensured a degree of economic leveling. Fear of such accusations encouraged those wealthy in cattle to be generous in their lending out of livestock to community members, perhaps softening jealous feelings that may have otherwise influenced the outcome of an umhlahlo. Also implicit within these stories is a criticism of Zulu kings unable to protect subjects from unruly chiefs and unscrupulous healers. The abuse of others by chiefs and healers seemed to increase during the Zulu civil wars and with the decline of the kingdom after 1879. Kings, who were already wealthy, were rarely accused of using healers for their own economic gains.

Europeans often suspected and accused Zulu kings and chiefs of rigging the umhlahlo in such a way that they could conveniently eliminate political opponents or enemies by ensuring they would be "smelt out" as umthakathis. Their concerns, however, focused more on the role of chiefs in Natal from the 1860s onwards. Healers were deemed instrumental in legitimating acts of political elimination and accused of being "political engines in the hands of the chief." Natal colonial administrator Theophilus Shepstone claimed in 1881 that "if the chief fears a strong member of his tribe, it is only necessary for him to induce the witchdoctor to point him out publicly as guilty of witchcraft to accomplish his ruin."[68] These accusations presumed that healers had little autonomy but were securely within the pockets of African rulers. While it is difficult to judge the exact degree of autonomy that healers exercised, particularly in the early to mid-nineteenth century, it can be argued that healers did not fit into the neat hierarchical model imagined by Europeans. Government administrators had correctly recognized healers as important political actors within Zulu society, both in the Zulu kingdom and in Natal. Their model, however, did not consider the more democratic mechanisms of control that existed within the umhlahlo circle or the fact that relations between healers and rulers could sometimes be quite contentious. Nevertheless such analysis informed colonial reaction to African healers and rulers.

Outsiders' perceptions notwithstanding, political opposition could not be eliminated by using the umhlahlo alone, as the ceremony's structure provided checks on the power of healers and rulers. These checks are evident in the participatory nature of the *bula*-circle, where participants' contributions were vital to the discovery of the umthakathi. In this way, participants, usually important men and women of a chiefdom or nation (depending at which level it was held), wielded much political leverage. It was the participants of the umhlahlo who revealed the identity of the umthakathi; the sangoma (who was often from outside the community) mostly confirmed and sanctioned public consensus. The assumption that political opponents could be eliminated via the umhlahlo neglected the fact that participants could also denounce an isangoma's findings. An innocent person who was "smelt out," for instance, could ask a healer to pay reparations of cattle for libel. Furthermore, many isangomas attended umhlahlos with armed followers who protected healers from being stabbed by those they accused and enabled them to run away after making accusations.[69] On the other hand, a renowned healer much respected and revered by the nation or chiefdom could influence the outcome of an umhlahlo, and in this sense perhaps Africans and European colonists were both right in their skepticism of this process. What they over looked was the fact that healers themselves could act as critics of the Zulu political system.

While healers were needed to initiate, protect, and empower a chief or king, they could also present a potential political threat, and thus various precautions were taken to keep healers firmly under the control of the Zulu kings and chiefs. Endowed with the blessings of the ancestors, the powers of divination, and knowledge of powerful medicines, healers could have potentially developed an independent base from which to launch complaints. At various times in the Zulu kingdom, isangomas seem to have shown some signs of independence from the king. This was never to the extent of neighboring Xhosa and Shona healers, however, who successfully claimed a power base independent of their rulers. In the stories told of healers during the early period of the kingdom under Tshaka's rule, isangomas offered critiques by smelling out members of the royal house.[70] As Lady Barker relates Zulu lore some sixty years later, healers "were in the habit of denouncing as witches, or rather wizards, one after the other of the King's ministers and chieftains."[71] In one story about this period, a more knowledgeable healer indirectly critiques Tshaka by reprimanding another seemingly less reliable healer for giving Tshaka medicine that would cause him to kill his own people. Upon hearing this Tshaka allegedly ordered the killing of the more experienced healer.[72] The umlozikazana posed a particular potential threat to local and national rulers. Unlike healers who depended upon the consensus of the bula-circles, many of these healers who channeled the voices of ancestors operated alone. In one story, Botshobana ka Sibaxa of the Tembu, reputed to be a great umlozikazana, claimed to possess all the Tembu chiefs within him. Botshobana's umlozi were heard to demand the slaughtering of beasts, a sacrifice often demanded by the ancestors. While some reacted with skepticism, other persons readily complied. The direct approach of the umlozis was apparently quite convincing and made umlozikazanas quite popular. Such autonomy, however, could be used to critique the king. It is thus not surprising that kings such as Tshaka forbade healers from using this innovation. (Dingane showed greater tolerance and allowed umlozikazanas to thrive at the homestead level but excluded them from the umhlahlo.)[73] In all of these stories healers are powerful, either in their ability to unearth the truth and protect the common person or in their ability to manipulate others and work on behalf of those who are dangerous— be they kings or chiefs.

In addition to their potential to act as political critics, healers also posed a threat by their very knowledge of and proximity to their rulers. If a healer switched political alliances, as Tshaka's healers claimed they were when they approached Zwide, their knowledge of chieftain/kingship medicines, as well as their direct access to a ruler's body dirt and personal items and the nation's inkatha, could prove invaluable to an adversary. On a practical level, healers had access to strategic economic and military intelligence useful to enemies. Even within the nation, fear of a healer's political allegiance to one ruler over another led Dingane to question the loyalty of Tshaka's doctors, whom he had hired to bring rain. When lightening struck near the royal isigodhlo Dingane claimed that the doctors had a "political motive" and purportedly had them killed.[74]

The consolidation of state power in an area where muthi, healers, and political power were so intertwined necessitated close ties between rulers and their healers. Various strategies were employed to limit the power and influence of certain healers. As previously mentioned rulers often asserted their authority by promoting themselves as the most powerful healer of the land. Another means of controlling healers was to limit who was allowed to practice.[75] Ukokela, a steward of Mpande who already retained an important position within the kingdom, started to undergo the *ukutwasa* illness exhibited by persons selected by their ancestors to become isangomas. When the king allegedly discovered Ukokela's ailment, he confiscated all of his cattle, successfully curing Ukokela of his affliction.[76] In an effort to limit the number and powers of healers, Tshaka was said to have concocted a plan that entailed his smearing blood on his isigodhlo to help discern genuine isangomas with clairvoyant powers from those less genuine. On gathering together the isangomas of the nation, he told them that an umthakathi had bewitched him by pouring the blood of a beast on the door of his isigodhlo, and he asked them to name the culprit. Of the group present only two isangomas claimed that it was not the work of an umthakathi but "done only by the heavens above" — meaning the king himself. Tshaka then reputedly executed all the isangomas except the two who knew how to "bhula."[77] "After this," Ndukwana declared, "there was no initiating of diviners, and no running about crying. Women put on the topknot, and men put on the head ring, and they became ordinary people."[78] While the veracity of this story is questionable, as different versions have different numbers of surviving isangomas, it does demonstrate the tension between healers and their leaders and the attempts of rulers to curtail healer's influences. Among the Sotho, a similar story exists in which chief Mabulane tested his doctors by putting fruit in his mouth and claiming he was ill. The one doctor who realized Mabulane's ruse was then declared the only true doctor and the others returned to their homes.[79] Reflecting on Tshaka's hoax, Arthur Shepstone suggested that this was "not due to a laudable desire to exterminate those who encouraged it (witchcraft) but rather a fear of their power over people which might have at any time been directed against himself."[80]

Among the Zulu, this testing and supposed killing of healers did not deter a new class of isangomas from arising. In fact their numbers increased so dramatically during the reigns of Mpande and Cetshwayo that both kings sought to control their numbers by instituting isangoma-only regiments. Isangomas had been previously exempt from military service since their work with powerful and dangerous muthi could affect ordinary soldiers.[81] The large number of male isangomas during this period may reflect the fact that this was one of the few means by which men could oppose military service and resist the rigidly prescribed gender roles of the period.[82] An isangoma-only regiment allowed the king to keep a close eye on problematic isangomas as well as acquire their labor and attempt to stem the tide of men avoiding military service. This *ibutho* of isangomas, however, did not fight in wars but did tend to the king's gardens, mend fences, and perform other work of the king's regiments.[83] Such a formation may also have removed isangomas from their communities during certain periods and thus have had the effect of reducing deaths from accusations of witchcraft.

Rev. Callaway also claimed that chiefs and kings sometimes employed assassination as a means to combat the potential threat of healers. Callaway went so far as to write in an explanatory footnote: "When a chief has obtained from the diviners all their medicines and information as to the mode of using the *isithundu*, it is said that he often orders them to be killed, lest they should use their sorcery against him."[84] Not only is this accusation unsupported by Stuart's respondents, it seems counterintuitive, as it would make recruiting healers for a chief or king's initiation difficult. Rather, it seems healers gained a certain prestige and other benefits from their close connection to a chief or king. A common Zulu refrain noted: "The doctor of the chief is a doctor indeed."[85] Even Callaway's own respondent tells how healers tempted new chiefs with powerful muthis in hopes of becoming their personal healers.[86]

CONCLUSIONS

As the stories of the Zulu kingdom demonstrate, muthi and healers played an important symbolic and practical role in achieving and maintaining political power in the nineteenth century. Given the militancy and uncertainty of the period, Zulu kings found it politically expedient to incorporate powerful healers into the ruling elite. Kings thus utilized healers' social and ritual powers and legitimacy to embolden and maintain their own power base; yet they also acted to curtail their power and autonomy. Historian John Laband observes that following Tshaka's killing of the isangomas, "the ritual authority of the Zulu kings went unchallenged, and their beneficent ancestral spirits were regarded as those looking after the interests of the whole nation."[87] Negotiating the tenuous relationship between healers and political leaders, however, was an ongoing process that different kings approached differently. Whereas Tshaka and Dingane were said to kill healers who threatened their rule, Mpande and Cetshwayo confiscated their property, conscripted them into the army, or lessened the impact of their verdicts by allowing the accused to seek refuge under the king. The Zulu state's dependence on healers and the limiting of their autonomy helps to explain, in part, why Zulu healers did not play an independent role in anticolonial movements. Among the neighboring Xhosa and Shona, who helped lead anticolonial movements, healers not only acted as the guardians of public health and well being but were endowed with power independent of their political rulers. Consequently when political rulers' authority and legitimacy weakened under the tightening noose of colonialism, charismatic healers emerged as the new political leaders and led anticolonial movements in southern Africa in the 1810s, 1850s, and 1890s.[88] When Zulu healers did play an important role in anticolonial movements, such as in the Anglo-Zulu War and in Bambatha's Rebellion in 1906, they did so at the behest of their rulers.[89] It was only after the fall and disintegration of the Zulu kingdom and weakening of the chiefs that healers again gained autonomy in the area. Stripped of the power of the Zulu state, however, their political power and legitimacy were greatly diminished.

In considering the autonomy of healers that accompanied colonialism, it is interesting to note how the use of human muthi increased within the colonial setting. White authors writing about this phenomenon referred to it as a superstitious custom long in practice among African chiefs to "gain an advantage over rival chiefs."[90] Fynn claimed that Zwide disemboweled Dingiswayo and drank his gall in order to obtain the strength and courage of his enemy, after which he rendered the body to fat. Fynn maintained that this practice was common and eventually spread to the Zulu kingdom from the areas north of Delagoa Bay.[91] Stories of murder for human muthi circulated widely in the twentieth century among the white community, particularly after a white man was murdered for war muthi during the unsuccessful Bambatha Uprising in 1906.[92] Among African respondents, the use of human muthi in the precolonial period was mentioned, but as three of Stuart's respondents told him: "We cannot declare it to be an invariable custom."[93] African stories regarding the use of human muthi differ from white ones. In one it was alleged that the Cube used human fat in the smithing of iron to make it pliable. In another, Malambule ka Sobuza, a royal Swazi refugee to the Zulu kingdom, was alleged to have dug up the body of King Dingane, which he rendered to fat and used to overcome a rival Swazi faction. In a third, Zibebu was suspected of treating his army during the Zulu civil wars of the 1880s with human muthi.[94] The term *inswelaboya,* used to describe a hairless individual who murdered to obtain flesh for medicines, is said to come from the Chunu chief Pakade ka Macingwane, who left the Zulu kingdom for Natal in 1838 or 1839. Pakade supposedly blamed the inswelaboyas to cover up his own indiscriminate killings. The made-up figure of the inswelaboya, however, seemed to have had real counterparts who attempted to harvest body parts and bodily fluids. In Natal, Madikane relayed two stories of attacks on boys in Pakade's district, one of which occurred in the late 1840s, but both of which involved suspicious attacks and pointed to the purpose of collecting human muthi.[95] In all the stories told by Africans at the turn of the twentieth century, it is outsiders within the kingdom, or Africans in Natal, who participated in this abhorrent practice; stories told by whites restrict the collection of human muthi to the empowering of chiefs over rivals during times of war.

The reality of muthi murders and their seeming spread among ordinary Africans in the late nineteenth and early twentieth centuries is evidenced by the court records.[96] This seems to have been the result of two phenomena directly related to colonialism rather than the continuance of "age old traditions." First, such practices should be seen as a reaction against the increasing social, economic, and political pressures that accompanied colonialism and land alienation. Desperate times called for desperate measures. If we are to believe the stories of human muthi from the nineteenth century, both African and white, we can observe an escalation in the obtaining of more and more powerful muthi. Second, as the Zulu kingdom and power of the chiefs came increasingly under attack, healers gained greater autonomy and certain practices previously restricted to the ruling elite gained currency among more ordinary Africans. Historian Rob Turrell traces muthi murders in early twentieth century Natal to chiefly rivalries, but maintains that muthi killings for ambitious individuals became evident by the 1920s.[97] The autonomy of healers was obviously gained at a much earlier date in the neighboring colony of Natal than in what would become Zululand. The next chapter discusses not only the impact of this autonomy on healers and healing practices but the impact of colonial legislation on the distinction between and practice of inyangas and isangomas.

The political connection forged between healers and their rulers during the early years of the Zulu kingdom and witnessed by British traders and settlers would in turn greatly affect the response of Natal colonists to African healers throughout the nineteenth century. The fact that stories of muthi and power as relayed in this chapter were still circulating at the end of the nineteenth century also helps to explain why colonial officials were keen to curtail the use of muthi and criminalize those who perpetuated its use. As we will see in the next chapter it was the anti-witchcraft function of African healers and their connections to hereditary chiefs within Natal that made them particularly powerful individuals and rather unpopular with missionaries and government officials. Even after Zulu-speaking communities and the Zulu kingdom were forcefully brought under British colonial rule and law, isangomas, at the urging of community members, continued to "sniff out" alleged witches, who then often met their death. As arbitrators of justice, healers represented the existence of a judicial and political system that interfered with the implementation of white rule. The more the British extended their political power within this region the more they realized their need to limit the influence of African healers.

PART II

Negotiating Tradition and Cultural Encounters in Natal and Zululand, 1830–1948

3 ⤳ Early African-White Encounters

Healers, Witchcraft, and Colonial Rule, 1830–91

IN THE SUMMER OF 1876, fourteen years after the criminalizing of African healers, Lady Barker wrote to her sister about a tea party she had held at her home in Pietermaritzburg. On her invitations she dubbed the party "Tea and Witches."[1] To this event she invited friends as well as the local African isangomas who were to act as entertainment for her guests. On the day of the event the isangomas, according to custom, came accompanied by a large number of local Africans. Not anticipating this presence, Barker seemed somewhat unnerved but reports that this "immense mob of shouting, singing Kaffirs" were as "docile and obedient as possible." At five o'clock Barker and her guests sipped tea and ate cake and biscuits on the veranda while five female isangomas "appeared in full official dress, walking along in a measured stately step, keeping time and tune to the chanting of a bodyguard of girls and women, who sung continuously, in a sort of undertone, a monotonous kind of march." After the women healers performed a mock witch-finding session, one of Barker's guests challenged the isangomas to discover what item he had recently lost. Using a bula-circle like the ones previously used to "smell-out" witches and criminals, one of the isangomas correctly announced that he had lost his pipe stem; Barker's guests seemed impressed. After the isangomas left, however, the guests concluded the evening with tales confirming for themselves the "savage" nature of Africans and the "wholesale massacre" of alleged witches before the English arrived and outlawed the practice of witch-finding.[2]

African healers, or "witchdoctors," as they were often called because of their reputation for pointing out witches, came to serve as the archetype of African superstition within European discourse. Drawn with wild eyes and often accompanied by snakes and a human skull full of frothing medicines, healers represented for the European and American imagination all that was "tribal," "superstitious," and "primitive" in Africa. At the 1939 World's Fair in New York, the quintessential celebration of modernity and technology, one exhibit featured the African doctor prominently in its "maze of superstition."[3] These images, which occasionally resurface in today's popular culture and helped Europe to rhetorically justify the colonizing of "savage" peoples, emerged in the Western world largely as the result of the memoirs of early travelers and settlers, as well as missionaries' and doctors' descriptions of African healers during the late nineteenth and early twentieth centuries.[4] Such conceptions were reiterated in local newspaper coverage of the times and later confirmed for many by early twentieth century anthropologists. Frustrated by Africans' initially low conversion rates to Christianity, missionaries blamed African healers for perpetuating African superstition. When appealing for money from their home congregations they invoked the image of the "witch-doctor" as an example of the many obstacles Christian missions had to surmount. These images had a profound effect not only on the ways in which Europeans and white South Africans viewed Africans, but on the ways in which they viewed themselves and their own colonial mission. Yet healers and ideas of African superstition served as more than a foil; they belie a complicated relationship between these two cultural groups.

More than a mere trope, superstition—particularly the belief in witchcraft—was perceived as a genuine liability and threat to colonial rule and missionary endeavors. Witchcraft proved a bane to both Africans and whites alike. African communities who suffered at the hands of witches (who brought illness, death, and misfortune) sought to discover and expose those who practiced it, while whites aimed to protect the accused and persecute accusers. Each saw their intervention as necessary, just, and preventing imminent death. This chapter examines some of the ways in which Africans and Europeans attempted to deal with these perceived threats. As the British endeavored to establish rule and law within the colony of Natal, Africans and Europeans sought to convince each other of the value of their tactics for combating this particular problem. This involved a great deal of negotiation on the part of both Africans and Europeans and ranged over time from amenable compromise, to open and hostile conflict, to the somewhat uneasy resolution we see at Lady Barker's party. Indeed this conflict led to perhaps the most enigmatic aspect of "Zulu medicine": the 1891 licensing of herbalists in the areas of Natal and (in 1895) Zululand.

While witchcraft played an important role in shaping African-European relations in the nineteenth century, medicine itself also proved an important means of opening and sometimes mending these intercultural relations. The medical "prowess" of Europeans led to the opening of trading and mission stations, while African medicine saved the lives of Europeans who lacked the experience or nostrums for healing local ailments. White doctors initially seemed to respect the work of African healers, seek their assistance, and even borrow their remedies.[5] By the early twentieth century, however, biomedical doctors grew increasingly skeptical of such healers and their abilities and began to claim that African healers were "unscientific" and "ineffective." The reasons for this transformation reflect changes in both colonial strategies and biomedicine during the nineteenth and early twentieth centuries. Likewise, this period saw the most radical shifts in the practice of African therapeutics. British legal changes forced African healers to practice medicine differently, in the ways in which people were both diagnosed and healed. Consequently this chapter shows how white rule influenced African medicine from an early period, particularly as the colonizers sought to reshape African ideas of witchcraft.

MEDICINE AS A GATEWAY OF ENCOUNTER

The exchange of African and western medicine and healing of patients across cultures occurred from the time of the first African and white encounters and continued over time within the Zulu kingdom and the colony of Natal. Such medical encounters also provided the opportunity for other types of exchanges. For instance, European and American missionaries seeking converts in Africa often employed the services of medical doctors. By healing the body, missionaries hoped to capture the hearts, minds, and souls of Africans as well. Not only was medicine used as a tool of the missionary endeavor in Africa,[6] it also provided a means to secure other secular and nonmedical objectives. This notion was grasped early on by European settlers and later adopted as policy by Natal legislators in the 1890s. Just as Zulu kings used medicine as a means of demonstrating their knowledge and power over ordinary Zulu citizens, healing the body became a means by which Europeans sought to demonstrate the superiority of western medicine and, by extension, western culture, civilization, and religion. Medicine became a potential gateway to changing African attitudes and knowledge of the body, illness, and misfortune. The process was, however, slow and uneven, and one which did not take root until colonialism had become well entrenched. During the first encounters of whites and Africans, however, medicine became an important means of establishing dialogue and trust, as well as a means of physical survival.

Illness often has a way of opening minds to alternative therapies; furthermore, Europeans and Africans exhibited an ideological receptivity that encouraged the use of each other's medicines in the early nineteenth century. Believing that nature provided local remedies for local diseases, European travelers, traders, and biomedical doctors within the colonies initially sought local cures for local ailments.[7] Likewise local African medical beliefs included the notion that strong medicines came from the farthest territories. The proverb *"imithi ikhendlwa kwabezizwe,"* or "Potent medicine is best got amongst aliens," underscores this belief.[8] Such thinking not only helped to initiate the use of foreign healers and medicine but sometimes resulted in tangible material and social benefits.

Henry Fynn, like many other early European settlers, had no formal medical training and allegedly gained his reputation as a healer within the Zulu kingdom by using common household remedies of Europe, such as chamomile, together with local therapies. He maintained that his medical aptitude, specifically his successful treatment of King Tshaka's stab wound, resulted in Tshaka's granting of land to the settlers at Port Natal.[9] Similarly, the Lutheran missionary Scheuder found his previously denied request to set up a mission station within the Zulu kingdom was granted after he successfully treated Mpande for his rheumatism and promised to provide future medical services for the king.[10] Clearly whites sensed that European medicines—folk or patented—were valued by Africans and could help smooth cross-cultural relations. For instance Captain King brings Tshaka European medicine on his initial visit, whereas Mr. Petersen attempts to impress Tshaka by showing him pills purportedly suited for all illness; likewise settlers Isaacs and Fynn both tell the often quoted story of Tshaka's vanity and demand for macassar oil with which to hide his graying hair.[11] These stories, whether exaggerated or not, tell us that medicine was an important means of opening relations between the African elite and Europeans. We can also see in writers such as Fynn and Isaacs that such stories constituted part of a growing colonial discourse that portrayed Africans as gullible and unscientific. Fynn tells his readers that Africans "supposed him to be some superior kind of doctor, possessed of all the magic powers requisite for causing and presenting rain and thunder."[12] But this same discourse is also meant to show the cleverness and guile of the white settlers. In the face of African superstition and Tshaka's irrationality, a Western reader's fear for Fynn or Isaacs's safety is assuaged by a contravening discourse of white rationality and settlers' ability to use African gullibility to their own advantage. While admonishing African superstition, Fynn also clearly notes and respects the more "rational" side of African medicine and readily adopts its use.

By itself, medicine could be readily adopted outside the cultural milieu from which it emerged and assume meanings not necessarily intended by those who originally produced or administered it. This was as true for Europeans as it was for Africans. When treating African patients, Fynn tells us that he conformed to certain African expectations and used both European and African therapeutics. While part of this involved a cultural performance—as Fynn tells us, "conceit can kill and conceit can cure"—saving lives and conforming to African cultural medical expectations were necessary to winning and keeping the favor of Tshaka. Yet the evidence also suggests that Fynn used African medicines to treat other traders and white settlers.[13] One such settler, William Bazley, tells James Stuart how Fynn had helped save Bazley's father's life in the 1850s after his father had suffered a severe leg wound. Fynn stopped the bleeding by balling up a number of spider webs and placing them over the wound, which was then bound up with a cloth. This European home remedy was combined with local herbs, which Fynn sent for and pounded into a poultice which he laid over the web. Bazley recalls that Fynn called these "kaffir medicines, some of which were very powerful." In another medical emergency, the Bazleys again called for Fynn's assistance, and he responded by sending "a native to a neighboring kraal for a particular drug."[14] In both instances, there was no stigma attached to using African medicines, and in fact their use was a necessity. Even at a much later period, in the 1870s, we see that biomedical doctors sampled local remedies; Campbell admits to using mimosa bark as a remedy for diarrhea and dysentery.[15] Like Fynn, Campbell observed local treatments by watching African practitioners in cases where several doctors had been called on.

Africans not only used European medicines and appropriated certain European remedies, they also used "European" substances not intended for healing purposes. During periods of crisis such as the feverish epidemics that raged along the coast of the Zulu kingdom in the early 1850s, Africans besieged missionaries like Scheuder for medicines.[16] Missionaries and white traders who offered and sold medicines helped to create a market for many patent medicines within the Zulu kingdom as well as Natal, and by the late nineteenth century many Africans in the countryside sought out these patent medicines by mail.[17] Africans also adopted European remedies readily available in the natural environment, such as horse manure, whose uses were learned from the Boers after 1838 and used to prevent childhood fevers,[18] as well as exotic flora that readily adapted to the South African soil. This included the two exotics mentioned in the Ngcobo trial: jalap from South America and male fern from Europe. Africans used these herbs for the same purposes as their European counterparts, and the Zulu term for jalap, *jalambu*, also shows some linguistic borrowing. Europeans had routinely used the male fern to counter tapeworm, but seem to have both adopted the local fern *inkomankoma* for this purpose and introduced the male fern on their arrival to South Africa. Africans also used European substances such as gunpowder and bluestone for medicinal purposes, particularly for counter irritation, as when these substances were rubbed into scarifications.[19] Later chemicals, such as bichromate of potash (used for developing film), became popular albeit dangerous ingredients of love medicines.[20] European remedies and materials sometimes took on alternative meanings not related to the physical cure, but this will be discussed in greater detail in chapter 4.

Divorced from doctors and institutions, European medicines were easily incorporated into African notions of medical care. Such European remedies and substances were generally seen as supplementary to rather than replacements for African medicine. Yet even when married to biomedical doctors and institutions, Africans judged biomedicine's efficacy based on African notions of proper medicines. Africans expected medicines to taste bitter, to stimulate the bowels or cause one to vomit, and to work quickly. Campbell writes of one occasion in the late 1800s when an African asked for a refund for his father's treatment: "How can it [Campbell's medicine] be doing good when his bowels have only been moved twice in four days." Though the quinine had stopped the chills and vomiting, Campbell complies with local expectations and prescribes calomel and jalap powder—a strong purgative.[21] Likewise Africans scoffed at biomedical treatment for syphilis, which required taking medicines for several years, deeming it ineffective because a cure could not be effected after a single bottle.[22] Even though Africans used Western medicines, their use was often sporadic. At certain times missionaries and district surgeons exclaimed the popularity of biomedicine; at other times they lamented African indifference and resistance to it. Some African wariness of biomedicine seems to have been encouraged by African healers from the late nineteenth to the mid-twentieth centuries as they began to perceive the ideological and commercial threat of biomedicine.[23] Throughout this period, however, Africans as a whole remained fairly consistent in their skepticism of biomedicine's invasive surgeries, hospitalizations, and quarantines and went to great extents to avoid such impositions.[24]

WHITE CHIEFS, AFRICAN SUBJECTS, AND THE PROBLEM OF WITCHCRAFT

Early European settlers were willing to accept that African medicines and healers could heal the corporeal body, and many whites in the nineteenth and twentieth centuries used isangomas, isanuses, and inyanga yokubulas to discover thieves and lost cattle. These same persons, however, winced at the activities of such practitioners when they claimed to be able to heal the body politic, particularly through their "sniffing out" of witches. Perhaps witchcraft represented for Europeans vestiges of their own superstitious past, but it was also something that distinguished them from their African counterparts in a European age of science and reason. Such variant views on witchcraft ensured that it would become an axis around which much cultural and political controversy would emerge in colonized areas. This can be seen in some of the first African-white encounters in south eastern Africa. The earliest white traders to Port Natal—Francis Farewell, Henry and William Fynn, Henry Ogle, John Cane, and Nathaniel Isaacs—initially acted as white chiefs over a small number of homesteads which then grew into small chiefdoms. Some married local women with whom they fathered children. Out of necessity they adopted African laws and customs in the running of their communities.[25] Prior to the implementation of British colonialism, such white chiefs found themselves in the awkward situation of attempting to dissuade African belief in witchcraft due to their own Eurocentric values while simultaneously maintaining political legitimacy among a population that insisted on strict and swift action against witches. Nathaniel Isaacs's story, as told in his popular *Travels and Adventures in Eastern Africa*, provides an interesting case in point.

In 1831, Isaacs agreed to hire an isangoma for a family who believed they had been bewitched. After an arduous process of price negotiation, a female healer came to the community with some forty armed men and performed a witch-finding session during which she fingered a suspect named Mattantarny.[26] She ordered that his residence be torn apart, in order to find the offending poisonous roots he used to bewitch the family. During the entire session, Isaacs and the Fynn brothers sought to discredit her, continually following and challenging her, until she threatened to disrobe to prove her integrity. Yet when they observed her helper drop a satchel of roots into Mattantarny's house they foiled her plan by snatching them and then showing them off to the crowd as a means to discredit her. What follows this incident is particularly interesting. Isaacs learns from his senate that it was well known within the community that Mattantarny was an umthakathi. He had allegedly been seen with an *imphaka* (a witch's familiar) and suspected of putting poisonous roots in people's calabashes. Surprisingly, Isaacs then brought charges of poisoning against Mattantarny. In the trial that followed a number of witnesses testified before Isaacs and his senior headmen. On this evidence, Isaacs and his "senate" found Mattantarny guilty of keeping an imphaka and poisoning people in an effort "to intimidate and affect" them.[27] Mattantarny was sentenced to death. In a note of triumph, Isaacs tells his readers, his actions "tended to remove a good deal of the superstition of my natives, and to impress them very strongly with the absurdity of their notions of charms and witchcraft."[28]

Given Isaacs's tendencies toward embellishment, one can assume a certain degree of exaggeration; nevertheless his story is instructive, particularly in what it suggests about European attitudes toward certain healers and toward implementing "customary" African law and how they perceived themselves. As newcomers, European traders and chiefs had relatively little power and could assert their prominence over Africans only by adhering to local customs and rules that predated their arrival. Establishing their position of authority meant mandating many decisions made by their "senators," or elder statesmen, regardless of their agreement or disagreement. Like the colonial administrators who followed and applied "customary law," these white chiefs used pre-existing African structures that helped them to establish authority while working to undermine those customs that challenged their power.

Isaacs's story not only highlights the means by which illness and witchcraft accusations were discovered and resolved during the late 1820s but also demonstrates the degree to which spectacle was an inherent part of the unveiling of alleged witches. From Isaacs's version of this story we find that both the community and the inyanga had already fingered Mattantarny as the umthakathi, yet the community expected the inyanga to vindicate their knowledge and produce tangible evidence. Part and parcel of this production was the inyanga's demonstration of her powers, both political and spiritual. She demonstrated this through an escort of armed men, by demanding cattle, beads, and cloth, and by verbally challenging the chief—Isaacs. All of this preceded what was to be the final and dramatic unveiling of the umthakhathi and the poisons used in the treacherous deed. Isaacs himself plays into this spectacle through his interactions with her but engages in his own show of publicly discrediting the inyanga and revealing her "evidence."

The palpable tension between Isaacs and the inyanga, and Isaacs's decision to sabotage her performance, represent not only a clash of two world views but also local tensions present between healers and rulers—be they white or African. She tested and demonstrated the desirability of her services by demanding payments and behaviors and by making pronouncements. Furthermore, she belittles his skepticism by offering to undress. Such an offer not only highlights his doubts of her powers, but insinuates that Isaacs does not respect the power of the ancestors. Likewise, her offer tapped into a gendered form of power and protest still used by southern African women today to shame men into proper behavior.[29] By shocking the crowd, she gained their support and confirmed her legitimacy. Her success thus further pressured Isaacs to comply with any other demands she made while simultaneously challenging his authority. Indeed, Isaacs's reaction to this healer and his decision to try Mattantarny on charges of "poisoning" rather than witchcraft indicate the potential political threat that such a healer posed. Given that social conventions necessitated intervention, Isaacs used his "senators," or the equivalent of the Zulu ibandla, as a means not only to circumvent and disempower a powerful individual in society, but also to administer justice in a more "transparent" form that enabled him to exercise more control and influence.

The European discourse of African superstition plays a particularly important role not only in denigrating African cultural ideas and practices, but also in obscuring the ironies of early white settlement in which Europeans were forced to adopt certain African beliefs and practices. Isaacs tells his readers that he agreed to hire the inyanga yokubulas because "I was elated with the hope that I might expose them, and thus extirpate the superstitious impressions of the ignorant by holding up the impostors to execration."[30] He portrays his intentions as seemingly noble and paternalistic. In effect, however, his reactions to this particular case seem not to dissuade African belief in witchcraft but only to discredit this particular healer. Isaacs himself based his decision on African notions of appropriate evidence: Mattantarny's feeding a witch's familiar, poisoning that was never witnessed or tested, and Mattantary's "looking guilty." Furthermore Isaacs and his senators inflicted the same exact punishment—death—as would have been administered in the Zulu kingdom at this time. The main difference, however, was that Isaacs contrasted his own method, which was supposedly rational, with African law, which he characterized as "Zoolacratic . . . the most incomprehensible government."[31]

While many isangomas, isanuses, and inyanga yokubulas may have been incorporated into or (in some cases) intimidated into working for the Zulu kingdom, the quasi-independent nature of the white chiefs may have enabled healers in these areas to exercise a degree of autonomy similar to that possessed by diviners prior to the consolidation of Tshaka's power. White chiefs, however, were not interested in co-opting these healers in the same ways that Zulu kings and chiefs were. Nevertheless, they did find it in their interests to control healers' power. Convention and people's desire for the intervention of healers in the eradication of harmful individuals from their community made it difficult for a white chief to decline a request for an umhlahlo. When colonial powers throughout Africa attempted to ban the ferreting out of witches by either healers or poison ordeals, they discovered that outright prohibition was an ineffective means of limiting healers' powers.[32] Thus, like Tshaka and the Sotho king Mabulane, perhaps Isaacs's only way to challenge the power of healers was by exposing them. When he did this, however, he was still left with an unresolved witchcraft accusation. While there may be similarities between African chiefs' and Isaacs's reaction to healers, Zulu kings and chiefs still needed to use healers—isangomas, inyanga yokubulas, umsutus, isanuses—to justify their rule and empower themselves politically, militarily, and spiritually in a way that white chiefs did not. As newcomers, these white chiefs were sensitive to any process that might either undermine or eliminate the need for their authority. As this example demonstrates, it was not the accusation of poisoning to which Isaacs objected but the process by which the culprit was discovered.

ESTABLISHING BRITISH COLONIAL RULE IN NATAL AND THE PROBLEM OF WITCHCRAFT

England's initial interest in Natal stemmed from concern over the potentially harmful effect on the Cape Colony of an unstable territory to its east[33] and fear on the part of commercial interests that another European power might gain access to a potential port on the southeastern African coast, thereby tapping into regional trade routes.[34] The British government decided to annex Natal in the early 1840s, after Boer presence along the border threatened to destabilize the Zulu Kingdom and in turn the settlement of Natal. The British sent in troops to push the Boers into the interior and officially annexed Natal in 1843. By 1845 they had instituted the Shepstonian system, a precursor to indirect rule that would later be applied throughout Britain's African colonies. This system of segregation and indirect rule through "native law" and "custom" was deemed the least costly and most effective means of avoiding confrontation with the Zulu kingdom.[35] At the same time, the power of the chiefs would be gradually restricted until Africans could be brought directly under European control.[36] As we shall see, the British establishment of colonial control over Natal led to more intensive interaction between the Zulu kingdom, the British, and Africans living in British territories. The better established the British became in Natal the more relations soured between the new colony and the Zulu kingdom, eventually leading to the Anglo-Zulu War of 1879.

Erroneously believing that all Africans lived under a chief on specified land, Theophilus Shepstone, diplomatic agent to the native tribes, aimed to more easily control the African population by rooting them to the land and a particular chief. During the first few years of British colonialism, Shepstone moved almost eighty thousand Africans onto specified African locations while marking other land for white ownership.[37] Organized into "tribes" under either a hereditary or an appointed chief, each group was to be overseen by a white magistrate and "cared for" by one or two missionaries.[38] Initially Shepstone estimated that "one-third, or at most one-half" of all Africans in the locations lived under hereditary chiefs, whereas the other chiefdoms created from fragments of other communities were artificially created units ruled by African assistants he trusted.[39] White farmers, however, soon complained that Africans had been given too much land, which enabled their economic independence and led to their unwillingness to labor on white farms. Given Britain's primary concern to maintain political stability, government officials sought other solutions to these labor problems such as recruiting African labor from Mozambique and later Indian indentured labor. This enabled Africans to remain in these locations, on unsurveyed Crown Land, and on unoccupied farms for a number of years.[40]

At the same time, the 1846–47 Natal Commission recommended that use of Roman law (used for the colony's white inhabitants) be dropped within these locations in favor of "African Customary Law." The main exceptions, however, were laws pertaining to African ideas of witchcraft.[41] Like the white chiefs who had preceded British colonialism in Natal, Shepstone appointed white magistrates to oversee local African institutions of governance. The magistrates then reported to him, the supreme chief or secretary for native affairs. "Customary Law" was based on the interpretation of elder Zulu-speaking men consulted by white magistrates or on a magistrate's "knowledge of the customs of the natives."[42] African chiefs were responsible for trying different types of civil cases. For a fee Africans could appeal any decision made by these headmen or chiefs to a white magistrate, or, after 1875, to the Native High Court.[43] The degree of judicial power chiefs held varied by magistrate. Both magistrates and Shepstone made decisions in an ad hoc manner that led to a number of inconsistencies and prompted Africans to protest in favor of a written code, which was finally compiled in 1875.[44]

Like the white chiefs before them, British colonial administrators worked to implement their own judicial system but also found themselves in the awkward position of having to intervene in cases of witchcraft, the existence of which they ostensibly denied. At the conclusion of an 1847 trial concerning the deaths of two alleged umthakathis, Lieutenant Governor West issued a warning, stating that the killing of "witches" would be dealt with severely by the new government.[45] Again the application of these laws was vague and left up to the discretion of individual magistrates. While there had been some recourse for those accused of witchcraft within the Zulu judicial system, these British laws empowered the accused in ways never before experienced. By opening up the courts to accused umthakathis and their families the British not only limited the persecution of alleged witches, but also enabled people to clear their names and retrieve property confiscated as punishment for witchcraft. A later case from 1878 shows not only that Nomatuse, a woman "smelt out" by an isangoma for causing her husband's illness by allegedly putting muthi in his beer, cleared her name but that the healer had to pay a £5 fine.[46] Because prosecutions lacked uniformity, much confusion resulted as they were implemented on the ground.

In 1850 this confusion manifested itself in tensions between two Natal magistrates who came down on opposite sides of a case of witchcraft. Chief Magedama of Impafana reported to Magistrate Peppercorne that a man named Filibi had been found guilty of killing three children by means of "witchcraft." In response, Peppercorne agreed to expel Filibi from Magedama's jurisdiction and sanction a fine in cattle, a relatively lenient punishment given the standards of the Zulu kingdom. When Filibi's cattle were taken, however, Filibi protested by appealing to another magistrate. In response to Filibi's complaint, Magistrate Cleghorn, along with 150 armed men, traveled to Filibi's district and seized 110 head of cattle from residents of the area.[47] This legal sanctioning of a witchcraft allegation by Peppercorne and Cleghorn's subsequent interference resulted in immediate attention from the colonial officials. Not only were both magistrates reprimanded by Shepstone, but in November of that year Shepstone issued a Zulu circular that was likely the first British mandate on witchcraft. The circular warned that anyone found killing another for "witchcraft or otherwise" would be subject to the death penalty.[48] The Filibi case resulted in two years of hearings and memoranda between the magistrates and Shepstone, debating the nature of witchcraft in African society and the proper colonial response.[49] In 1851, Shepstone wrote a memorandum that aimed to sensitize other magistrates to the realities of witchcraft accusations: "The question of witchcraft and its treatment by magistrates requires great caution and discrimination; on the one hand it is necessary to check it, on the other it is impossible to deny the existence of what the natives look upon as witchcraft, and which, whether it be the mere working upon the imagination or otherwise, does actually avail in causing both illness and death. I mean independent of the *administering of poison*, a not infrequent occurrence also."[50] While colonists speculated that many illnesses Africans attributed to witchcraft were psychogenic in nature, they also recognized that poisoning did occur. Distinguishing between the two, however, proved difficult. Evidence of poisonings became particularly apparent in the late nineteenth and early twentieth centuries, when ordinary people had more recourse to European poisons through cattle-dipping chemicals in the rural areas and arsenic and blue stone in the urban areas. While late-nineteenth-century government chemists complained that they could not recognize poisons of an organic nature, these common chemicals were traceable by chemists and are documented in court records.[51] Yet the dichotomy that Europeans drew between witchcraft and poisoning was not recognized by most Africans. Settlers like Isaacs, as well as colonial officials, thus replaced the idiom of witchcraft with that of poisoning, something that was abhorrent but understood within a European cosmology. Johan Struben, the resident magistrate of the Klip River division, argued along similar lines at the 1852 Natal Native Commission: "I would modify the punishment of witchcraft; I would not do away with it altogether, because it comprehends the charge of poisoning and the attempting to poison, which I believe to exist among them [Africans]."[52] An 1853 government notice referring to the 1850 circular attempted to persuade Africans against pursuing alleged umthakathis by giving graphic details of a man (or, to use local terminology, an umthakathi) who was tried and hung by the colonial government for *poisoning* his family.[53] Presumably this showed that the government also had ways of cracking down on alleged witches. Accordingly, Africans came to learn this idiomatic preference of the colonists and would later make their complaints to both colonial administrators and white doctors in terms of poisoning rather than witchcraft.

Magistrates on the front line of colonialism often had a nuanced understanding of local culture, and by utilizing a realpolitik they showed sensitivity to issues prone to disrupt relations between themselves, the state, and the local African community during these early years of colonialism. Another important observation expressed in this ongoing dialogue between white officials was the important social function played by witchcraft accusations. Peppercorne argued that witchcraft enabled communities to rid themselves of individuals who

> had become notoriously obnoxious to the chief and tribe by a continued course of misconduct, which could not be punished in detail by any native practice . . . They therefore allow these offenses to accumulate until some unforeseen or unusual death occurs, when the crime of poisoning is immediately charged upon the most notoriously obnoxious individual . . . They have in effect no other *criminal law*, and without it they could not purge their society of any bad members. To describe the practice as a mere gross superstition or that it is wholly included under our term "witchcraft," or to ridicule and hold it up as an inhuman practice, would shew a total misapprehension of the custom or rather law.[54]

In making decisions about whether or not to pursue a witchcraft accusation Peppercorne said he used two main strategies. He attempted to "ascertain that the chief has little or no individual interest in the punishment of the umtagati and that the punishment is inflicted as near as may be by general consent."[55] He stated that he tried not to interfere with the chiefs as he feared that this would make them "prejudicial to government."[56] Magistrates, though often paternalistic in their attitude towards Africans, frequently differed in their interpretation of necessary legislation from urban colonial administrators. At the same time, other colonial officials and missionaries were drawing political connections between healers and the power of hereditary chiefs, connections that would have implications for the way they approached healers.

Real or not, the existence of witchcraft threatened in subtle and not-so-subtle ways to undermine a sense of social order and, by extension, British colonial rule. Whereas colonial laws against witchcraft were initially drafted in response to the punishing of alleged umthakathis and focused on those who killed them, colonial officials came to perceive African healers themselves as a serious political threat to colonial interests. Unlike the indirect threat that healers posed to the white chiefs in Natal, healers in the neighboring Cape Colony led one of the more serious challenges to British colonialism in 1850–53. In western Xhosaland, under the influence and direction of a prophet named Mlanjeni, war-weary Xhosa-speakers united to challenge the dispossession of their land and foreign rule. Notions of witchcraft played an important part in this rebellion; people believed that witches had proliferated after the 1846–47 Xhosa War and during the ensuing drought and famine. Mlanjeni, a renowned witch-finder, not only planted poles to identify witches but claimed to cure them as well. Mlanjeni provided a major impetus for the 1850–53 war by assuring Xhosa chiefs and farm workers that their ancestors would empower them to fight the British. The word spread quickly, and by the beginning of the war in December 1850 almost every Xhosa worker had deserted his white employer.[57] This exodus was soon followed by five prophets who encouraged Africans in the Eastern Cape to kill their cattle and abstain from cultivation. They also predicted that black Russians would invade the colony and chase out the British.[58] The number of early to mid nineteenth century Xhosa uprisings in southeastern Africa—Nxele (Makanna), 1819; Mlanjeni, 1850–53; five prophets, 1855; and Nongquawuse, 1856)—vividly demonstrated to British colonists the ability of healers to mobilize public opinion and action for anticolonial purposes. Not surprisingly, British attention to witchcraft cases subsequently switched from concern for the alleged witches and punishment of those who killed them to a focus on and decision to outlaw healers themselves.

Though this switch was not reflected in legislation until 1862, the first proposal to limit the power of healers accompanied a discussion of decreasing the power of the chiefs during the 1852 Natal Native Commission.[59] The two came to be seen by colonial powers as intrinsically connected. Concern over the connection between powerful chiefs and healers had, in fact, been brought up by members investigating events in the Eastern Cape during the 1851 Select Committee on the Kafir Tribes. When asked whether the Xhosa prophet Mlenjeni "has with him the cordial sympathy of the people, or . . . is the mere tool of the chiefs," Major Bisset replied somewhat simplistically that he was "a mere tool of the chiefs."[60] In Natal colonial officials sought to implement their rule and improve security in the newly formed colony without replicating the mistakes that allowed such rebellions to arise in the Eastern Cape. Of particular concern to colonial officials was the power and influence enjoyed by the hereditary chiefs in contrast to appointed ones, who could be more easily controlled by the colonial government. Shepstone argued before the 1852 Natal Native Commission that despite deposing the hereditary chief Fodo and replacing him with his uncle, people had continued to follow Fodo, eventually forcing the government to reinstate him.[61] Such power led Jacobus Boshof, registrar of the district court, to state quite bluntly: "The power of these chiefs must be reduced; and we must do that now, or we shall find it to our cost hereafter, that two such Governments cannot exist. The one will destroy the other."[62] The very existence of British colonialism thus depended on finding the means by which to diminish the power of hereditary chiefs.

Beginning in 1852 colonists began voicing their suspicion that hereditary chiefs had maintained power and were eliminating political opposition through the mechanism of witchcraft accusations. Wesleyan missionary Reverend William David stated that witchcraft placed "in the hands of their chiefs, a power which, when Kafirs are in their independent state, is often made use of for the vilest and most tyrannical purposes."[63] Henry Fynn highlighted the role of healers in the process of these accusations. "He [a healer] is also the great lever by which the Chief exercises his power. . . . No chief who is actuated only by principles of justice could control a Kafir nation."[64] Such accusations gained ground after this period, being repeated and refined throughout Shepstone's tenure as secretary (1856–77). In 1864 Shepstone claimed that accusations of witchcraft were the "political engine" that could "overthrow the most powerful subject."[65] Almost twenty years later, in 1881, Shepstone was more explicit but still voicing the same concerns: "With independent native tribes the witchdoctor becomes a political engine in the hands of the chief. For instance, if the chief fears a strong member of his tribe, it is only necessary for him to induce the witchdoctor to point him out publicly as guilty of witchcraft to accomplish his ruin."[66] By 1892, he argued in retrospect: "Take away the engine and nothing will be left to lean upon but the power of the government."[67]

As these statements reveal, the decision to criminalize healers went beyond moral repulsion for witchcraft; it reflected colonial fear of the potential political power of healers, in their ability both to lead anticolonial movements and to uphold the power of the chiefs. The problem of hereditary office was confirmed in 1856–57, when two chiefs circumvented Natal law by administering their own punishments on criminal cases, one of which involved witchcraft. In addition they refused to account for their actions before a magistrate. In response, Shepstone disbanded and dispossessed the lands of Chief Matshana and Chief Sidoi, actions that served as an example to other hereditary chiefs who may have been tempted to challenge the colonial order.[68] These local challenges, as well as changes occurring in the neighboring Zulu kingdom and colonial presumptions about witchcraft and healers, most likely occupied the forefront of colonial official thinking when the British decided to ban African healers in the 1860s.

In October 1862, at the lieutenant governor's request, the secretary for native affairs sent all resident magistrates a circular that prohibited healers and rainmakers from practicing their skills within the colony.[69] It is not clear how well magistrates complied with this, the first law that outlawed the practice of healers themselves, rather than just the killing of alleged witches. Legally, prosecution was required but punishment was not specified. Archival records show that government prosecution of healers began taking place only in 1864.[70] Yet prosecutions of healers seemed to increase after 1875, probably due to the new 1875 written code of "Native Law." In Weenen, the district that recorded the largest number of people punished under these laws, fourteen of the twenty-two cases took place between 1875 and 1877.[71] Punishments, however, were inconsistent and ranged from fines of £2 to £10, with the possible inclusion of hard labor for three to six months. In one case of multiple convictions, a rain doctor had to serve six months of hard labor and forfeit all his cattle. In others, women erroneously categorized as practicing "witchcraft" were punished by having their hair cut off.[72] Interestingly Cetshwayo claimed to use a similar punishment by having healers' "hanging hair cut off."[73] Given that isangomas/isanuses customarily wore an *umyeko*, or twisted shoulder-length locks, which distinguished their occupation, such a punishment most likely shamed healers by making them metaphorically and literally indistinguishable from an ordinary person.

The same 1862 circular that had criminalized the practice of healers and rainmakers in Natal also forbade their entrance into the Zulu kingdom. Such actions presumably reflected the government's understanding of the political support such healers could provide in the neighboring kingdom and how events there concerned the well-being of the colony. A civil war had been waging in the Zulu kingdom from 1856 to 1860, and in the year previous to the circular Shepstone had assured Mpande that he would recognize Cetshwayo as his successor. The circular thus represents a means by which Shepstone sought to influence and limit the political and military power of the rival chiefs in Zululand as much as he sought to limit the power of his own hereditary chiefs in Natal.[74] Given the connections that colonists drew between the power of healers to help eliminate political opponents and the fact that they were the ones responsible for doctoring the armies for wars or faction fights, it seemed politically apt to again target healers and limit their control as one means of stabilizing events in the colony. This explanation gains credibility when one considers that the next circular concerning prosecution for witchcraft crimes was posted in June 1878 during the buildup to the Anglo-Zulu War. The 1878 circular prohibited the very consultation of "imposters [isangomas and isanuses]."[75] Given that people continued to consult healers, albeit in increasingly clandestine ways, presumably such legislation aimed at undermining popular support (monetary and otherwise) for such healers. Even though Shepstone had appointed many of the "chiefs" throughout Natal, the loyalty of hereditary chiefs remained questionable. Colonialists thus feared that with an impending war healers might not only empower the chiefs to challenge the Natal administration but also align themselves with the Zulu nation. In the 1880s the administration expanded and specified punishment for both healers and those who consulted them. The 1883 Natal Penal Code punished persons making allegations of witchcraft by a fine of forty shillings or fourteen days' imprisonment, and the "habitual *isanuse*" could be punished by two years' imprisonment.[76] This law was expanded to Zululand in 1887 under the Zululand Proclamation.[77]

Even after Zulu-speaking communities and later the Zulu kingdom (after 1879) were forcefully brought under British colonial rule and law, isangomas, at the urging of community members, continued to "sniff out" alleged witches who might then meet their death, be banished, or have their cattle confiscated. As arbiters of justice, healers represented the existence of a judicial and political system that strengthened chiefs and interfered with the implementation of white rule. The further the British extended their political power within this region the more they realized the need to limit the influence of healers and consequently passed more laws and specific punishments. Like the missionaries (discussed below) who became aware of the need to challenge and change the foundation of African thought at an earlier period and recognized the role that healers played in perpetuating African religious and cosmological ideas, British colonialists recognized the need to educate and convert Africans to a European knowledge system. In desiring to sell its cultural superiority and its idea of rational government and justice, they too came to realize that healers were in many ways the gatekeepers of local knowledge and governance. For all these reasons, the colonial government targeted healers, but particularly those who used supernatural powers, as a strategy for implementing British rule.

WHITE MISSIONARIES AND CONVERTING WITCHES

Instrumental to the British plan of establishing colonial rule was the role of missionaries. In many ways missionaries laid the groundwork for colonialism, providing knowledge of and contacts with Africans rulers in the precolonial period and often serving as linguistic and sometimes cultural translators. Christianity introduced far more than religion. Conversion meant adopting the cultural tenets of the converter. This included different ideas about gender, community, law, and what constituted proper behavior. As Lord Grey, secretary of state for war and the colonies, stated in 1847, "It is obviously impossible to expect these barbarous tribes to become civilized unless they are converted to Christianity."[78] Natal had one of the highest ratios of European missionaries to the indigenous residents in Africa in the nineteenth century.[79] In the same way that colonists hoped to use missionaries as a tool of empire, missionaries relied on colonial powers to implement their own agendas. Colonial legislators and missionaries, united in their repugnance of witchcraft, both saw the role of healers and African beliefs regarding health and illness as the root problems in dealing with African populations.

Mission stations were encouraged by the colonial state, and local African chiefs in Natal, much like the Zulu king, asked missionaries to reside on their land in hopes of benefiting from the non-religious services that mission stations offered. Of primary importance was a missionary's ability to function as a mediator between an African chief and white authorities. Secondary benefits included instruction in reading and writing, access to new technological knowledge, and use of medical services. Some of these skills became particularly valuable as Africans were further incorporated into the cash economy of the colony.[80] African rulers, however, took on a certain risk in accepting missionaries as these practical benefits could be outweighed if too many people converted and assimilated to Christian and western knowledge systems. Such assimilated persons were unlikely to accept the legitimacy of hereditary leadership, which based its power in part on the will of ancestors and strong muthi. It is for this reason that converts within the Zulu kingdom were accused of leaving the king.[81] Furthermore, as Paul Landau and Vaughan both aptly describe, missionaries introduced different conceptions of the world, ones that rested on individual and bodily orientations rather than community and social ones. Ultimately such ideas could undermine the power of witches and ancestors, whose actions rested on understanding an individual's interactions and responsibilities in relation to the community.[82]

Throughout the continent, however, missionaries initially had a hard time converting Africans to Christianity. In southeastern Africa, it was only after the devastating droughts, pestilence, and rinderpest epidemics of the late nineteenth century, combined with the tightening coils of colonialism, that Africans began to seek out new answers and convert to Christianity in larger numbers. Often the first converts to the mission stations were people who had been outcasts within the community. These included political refugees, women escaping unwanted marriages, and those accused of witchcraft or other crimes.[83] Locals, however, sometimes retaliated against those converting. Colonial administrator Robert Samuel reflects in 1929 about two women forcibly taken from his father's mission station in the Mhlatuze Valley. In one case the woman had been accused of witchcraft, and his father supposedly struggled unsuccessfully over the space of a mile to prevent her from being taken by group of armed men.[84] At the turn of the twentieth century, Zulu trader Crosby observed, "A native the day he joins the mission is looked upon as an outcaste from the tribe and on the Englishman's side."[85] During the early years, living at a mission station was akin to becoming kinless, as converts were often cut off from relatives and ancestors. Inevitably missionaries became surrogate families, even expecting future husbands to consult them and pay bride wealth for women at their stations.[86] The missions' harboring of and concern for alleged umthakathis; their collusion with the colonial government; their rules against polygamy and drinking *utwyala*, or homebrewed beer; and their espousal of a strange white god all gave rise to a great deal of African suspicion. Furthermore missionaries engendered resentment by trying to use natural disasters as opportunities for conversion. Astrup reports in 1896 that a reluctant group of women allowed him to pray for rain during a period of prolonged drought, after which he excoriated them, claiming that "God's punishment was upon them because they did not convert."[87] By this period rainmakers had been criminalized, though desperate people most likely continued to consult them. Yet the willingness of missionaries such as Astrup to use drought to prove the power of Christianity ultimately led to comparisons between the two powers, which sometimes backfired. As one of Stuart's respondents concludes, "The kolwas [African Christians] pray for rain and we attend their services and pray and pray with them, but rain does not come. . . . Formerly rain, when the season was dry, would be prayed for to our national amadhlozi, then the rain would come down. Hence we conclude that this former method, as regards rain making, was more effective than that followed by the kolwas."[88] The combination of all these factors, as well as the criminalization of healers, led another respondent to correctly speculate, "Missionaries are endeavoring to break down the native *uhlanga* [way of life] and destroy the power of the chiefs."[89]

Missionaries grasped early on that they were in direct competition with healers, whom they deemed the main agents of "non-Christian" beliefs. In many ways the position of the African healer paralleled that of the Christian minister; both served as conduits to powerful unseen forces. Likewise, ministers realized that African healers helped reaffirm the power of ancestors every time illness occurred. This is perhaps why Scheuder in 1851 attempted to dissuade Mpande from using local remedies, claiming, "It would be harmful rather than helpful to combine two so vastly different means of healing as those of the whites and those of his doctors."[90] Clearly this tension and incompatibility of systems was recognized by Africans and healers as well. While Africans might seek missionaries out for their medicine, conversion was another matter altogether. In 1851, "Veritas" complained that African families sometimes used healers to win back Christian converts, inducing converts "to say that they were in a state of madness when they embraced Christianity."[91] In 1891, Rev. Tyler alleged that an African healer caused converts to flee a Natal mission station by telling them that they would die if they remained there.[92] Reflecting on these continued tensions in the early twentieth century, Rev. Feyling of Mount Tabor Mission Station in Zululand writes, "In my opinion the real rulers of the natives are the witch doctors. This one thing I am sure of, that as long as witchcraft and witch doctors govern the hearts and minds of these people, there can be no real progress religiously, morally or intellectually."[93] Likewise, Rev. Hawkins of Lansdown Mission Station in Zululand argued, "Both the herbalists and witch doctors are the greatest hindrances to the Gospel and uplift of the people for they literally have the Zulus completely subject to them."[94]

African conversion to Christianity, however, did not always deter or limit the power of African cosmology. To the frustration of missionaries, many converts continued to use African medicine, seek rainmakers, and use love potions.[95] Other Christian converts continued to visit isangomas and inyangas, and accusations of witchcraft were often rife at the mission stations.[96] Witchcraft accusations proved as problematic for missionaries as they had for white chiefs and magistrates. Sanctioning a conviction potentially meant undermining their own work, yet denying such accusations put strains on a tenuous relationship where conversions were few and relations with non-Christians were sometimes openly hostile. Missionaries thus mulled over "witchcraft" accusations and in some cases agreed with the conviction.[97] Among the *kholwa*, the African Christian elite that developed at the end of the nineteenth century, there also continued a strong tradition of consulting diviners, using muthi, and fearing witchcraft. Missionaries consequently concluded that tailoring the power and prestige of the isangomas, isanuses, and inyangas to suit their own ends would make room for their spread of the gospel. Missionaries tackled this project in various ways; some directly challenged healers by providing medical care, while others put pressure on the government to limit healers' power. Either way missionaries proved great allies in the colonial endeavor to eradicate both the problem of witchcraft and the healers who perpetuated belief in it.

AFRICAN RESPONSES TO COLONIALISM
AND THE WITCHCRAFT PROBLEM

The nineteenth century had dramatically changed the means and mode of production for most Africans living in Natal and later in the Zulu kingdom. Their rapid incorporation into a capitalist economy dominated by whites led to increased tensions and severely weakened African social and political institutions. To secure African labor after 1854, the colony demanded that all incoming African migrants complete three years of compulsory wage service. Given the large influx of refugees from the Zulu Civil War, this amounted to a significant amount of labor.[98] Despite colonial efforts, however, many Africans refused wage labor, causing Natal to look outside the colony, both in southern Africa and in India, for a stable workforce. After 1875, Natal downsized African reserves to limit African agricultural output and competition against white farmers, thereby "encouraging" Africans to seek work on Natal farms.[99] In response to increased colonial intervention and the demands of South Africa's new mining and urban centers, as well as the drought and rinderpest epidemics of the 1880s and '90s, Africans in Natal and the former Zulu kingdom began entering the labor market in larger numbers by the turn of the century.

Colonialism brought new economic, social, and physical challenges that often worked together in a synergistic way to compound people's increasing health problems and the fear of witches. Not only did colonialism introduce new ailments, it destroyed local public health systems through the criminalization of healers and ecological controls by determining the movement and placement of people. Likewise urbanization and migrant labor fueled the rapid spread of new and old diseases. Colonialism had also created conditions for malnutrition and disease through the imposition of cash crops, forced labor, migration, taxation, colonial legislation, land dispossession, urban planning, and industrialization. Colonial interventions combined with ecological strains led to a prevalence of disease and a growing number of witchcraft accusations.[100] While the increase in disease itself may have led to increased perceptions of witchcraft, the stress and uncertainty of the period, as well as increased social mixing of people, all tended to add to people's anxiety about witchcraft.

Like Africans throughout the continent who experienced colonialism, many in Natal blamed the increase of illness and witches on whites whose laws and missionaries seemed to protect and side with the umthakathis. Thus when Lieutenant Governor West warned against the killing of alleged witches in 1847, he confirmed many Africans' suspicions of an alliance between white people and witches,[101] though whites themselves never seemed to be suspected of witchcraft. Through a messenger to the 1852 Natal Native Commission, Chief Pakadi implored, "The Government does not believe in the efficacy of witchcraft to kill people, and the Abatakati rejoice at this, because they see they can perpetrate their designs without molestation, and under its protection. . . . The government must not think we eat up Abatakati because we want their cattle, it is not so; we would gladly surrender the cattle into the hands of the Government provided it would forfeit them, and not give them back to the claimants."[102] Pakadi's statement not only addresses the suspicion that witchcraft charges were used solely to benefit the chief economically but offers practical solutions. Thirty years later this problem of witchcraft remained, leading another chief to speak more directly to the failure of colonial policy before the Natal Native Commission in 1881: "[The] Abatagathi have increased in number, because the white people don't believe in their existence and no means are taken by the Government to suppress them."[103] Likewise another chief testified, "We complain that [the] Government entirely disbelieves in the existence of abatagathi, though we suffer very much from their evil practices."[104] These concerns continued to reverberate throughout the late nineteenth and early twentieth centuries. Mkando attempted to persuade Stuart, as a colonial official, of the illogical response of the colonial government in banning healers: "White people are wrong in saying 'Did you see him [umthakathi]?' They give a lease to *abathakati*."[105] Given that umthakathis often inflicted illness by disguising themselves, using a familiar, or merely by spitting muthi in the direction of a victim's home, the only means of revealing a witch would be through the expertise of an isangoma. By protecting umthakathis, the most feared and hated persons in the community, the British did little to win African support for white governance.

Similar complaints were made about the criminalization of rain doctors. The coincidence of droughts with the passage of laws banning the killing of witches and practice of healers led Africans to make a direct connection between their misfortune and white rule. Before the 1881 commission, Chief Domba of the Amabele stated, "The Natives are perfectly loyal, but when a drought appears they complain that the Government won't allow the employment of rain doctors."[106] Africans also linked the problem of drought to witchcraft, assuming umthakathis had placed rain plugs into the land to prevent the rain from falling. When appearing before government commissions or talking with Stuart, many sought to negotiate with the government and convince them to decriminalize the practice of healers. Umganu, chief of the Tembu, told the 1881 Natal Native Commission, "I regret that Government has put a stop to rain-makers' practices. There are still abatagathi among my people, and I should like to have the power to drive these away."[107] Bikwayo, another respondent of Stuart's, attempted to bargain by arguing that rain was not good just for Africans but for whites as well, and that Africans would provide this service for free. "We cannot understand this as the Government would benefit because their trees would grow, and people would be free from hardship."[108] In these ways colonial rule seemed to defy common sense, forcing Africans to defend themselves against the onslaught of witches while also trying to avoid prosecution from an unsympathetic white government.

In the face of repressive colonial laws and punishments regarding the killing and unveiling of witches, isangomas began to practice their occupation clandestinely. The bula-circles that had been conducted openly in the presence of the entire African community and sanctioned by the chief largely disappeared with their criminalization in Natal in the 1860s. Rather than suppressing witchcraft belief, however, colonists merely changed the way in which people diagnosed illnesses. Those cases that involved suspicions of witchcraft were now held in seclusion, beyond the gaze of colonists and sometimes chiefs.[109] Before the 1881 Natal Native Commission, Africans insisted that divining was restricted to curing illness, not discovering witches: "There is no witchcraft as formerly by smelling out; it is confined to smelling out diseases."[110] In hopes of decreasing the power of chiefs, colonialism not only forced this practice underground but opened up room for abuse in the system by making healers less dependent on chiefs, who had previously curtailed healers' social and political power. Gama, a Stuart informant, testified to the national effects of similar laws in neighboring Swaziland. The criminalization of healers, he argued, had caused the Swazi king to lose control over them and the punishment of alleged umthakathis. This had empowered local Swazi chiefs, who utilized healers to divide the wealth of accused umthakathis among themselves. This, he claimed, "is how the isinduna (headmen) of the Ngwane acquired their wealth."[111] Something very similar seems to have occurred among healers in Natal and Zululand, as evidenced by the rise of muthi murders in the twentieth century.

While the criminalization of healers brought major changes to the practice of African therapeutics in Natal, living in Natal itself brought about increased intercultural encounters, not only with the white community but with a variety of different African communities that had migrated to Natal for work. Within Natal the appearance of new apparitions and forms of medical practice emerged in the second half of the nineteenth century long before materializing in the Zulu kingdom. In 1868 Callaway wrote about the *tokoloshe*, a Xhosa apparition said to be short and hairy and prefer the company of women. Callaway's respondent claimed that tokoloshes did not exist among the Zulu but that he had heard of Zulus who had seen this creature.[112] By 1891 the tokoloshe had come to Natal, and in the first half of the twentieth century was allegedly terrorizing women; its fat was also being sold as a powerful muthi.[113] Importantly, the Sotho also introduced new forms of divination that could be practiced more clandestinely than the bula-circles. These included divining bones and sticks that could deduce the origins of illness or misfortune, but they do not appear in the historical record of Natal until 1870 with Callaway's *The Religious System of the AmaZulu*. The sticks, or *umabukula*, allegedly jumped about independent of their owner and pointed to the area of illness; while such healers were highly prized, very few seemed to practice this art. More popular was the use of divining bones carried in a satchel and thrown out onto the ground, where the significance of their placement was read by a diviner. Divining bones' popularity most likely emerged due to the criminalization of the bula-circles, the fact that healers could be taught to read the bones, and the convenience and secrecy of being able to perform the divination at the homestead level. It is possible that this form of divination emerged earlier. Callaway, who lived near Richmond in Natal from the late 1850s to 1873, does not mention that this form of divination is particularly new. Another source from 1879, however, mentions that bone throwing is less popular amongst the Zulu than the Sotho. A photo postcard from the early 1880s shows an inyanga and two isangomas studying the layout of bones thrown on a mat (see page 8).[114]

In addition to changes in the practices and tools of healers, colonial pressures also seem to have effected the gender composition of healers. Evidence from both the Zulu kingdom and the early Natal colony indicate that healers could be either men or women, though no sources mention a predominance of one sex over the other.[115] The only exception occurred within the Zulu kingdom in the 1870s when male initiates and isangomas had been so numerous that Mpande and Cetswayo had instituted an isangoma-only army regiment, named Kandempemvu.[116] By the early twentieth century, however, the practice of diviners had become overwhelmingly dominated by women.[117] Bryant indicates that this switch to a predominately female occupation occurred in the late nineteenth century. As African men left the rural areas to work for white employers, women began taking up this occupation with greater frequency and their numbers became more noticeable. In Natal men were under increasing pressure to join the colonial labor force, and could thwart the process of initiation (ukutwasa) by taking certain medicines and making sacrifices to one's ancestors.[118] While evidence of this can be seen from the 1870s, it seems to have occurred with greater frequency in the twentieth century as men wishing to engage in wage-labor or Christian converts sought to prevent ukutwasa. Despite the criminalization of isangomas, this practice may have represented one means by which women could gain access to the cash economy from which they were largely excluded. Judith Gussler also raises the interesting possibility of a connection between ukutwasa initiation and the illness pellagra, which is brought about by malnutrition and seems to mimic the symptoms of ukutwasa.[119] Indeed given the ecological challenges of the late nineteenth century it would not be surprising that women who often ate last would have suffered from this disease. Consequently, nineteenth-century African women, deemed legal minors by colonial law and excluded from the cash economy, may have been the ones who embraced and reinvented the practice of the isangoma within Natal. These women not only used this occupation to empower themselves socially and economically, but played a crucial role in maintaining and negotiating local ideas of health and healing throughout the turbulent nineteenth and twentieth centuries.

Though shortly outside the time period of this chapter, the problems of the late nineteenth and early twentieth centuries only furthered African concerns regarding the efficacy of "traditional" means of improving health and well-being. During the 1890s, when Natal Africans not only suffered from drought and the rinderpest epidemic, but as many more were absorbed into the ranks of migrant labor working in the gold mines of Johannesburg, Africans claimed that the amadhlozi had turned their backs, leaving Africans to fend for themselves. In 1902, Mkando, an inyanga, lamented:

> Nowadays a man dies even when it is said by *izinnyanga* that *amadhlozi* are the cause of the illness. Formerly if a person fainted suddenly, had water poured over him, and messengers were dispatched to *izinnyanga*, whilst they were away the person would come back to consciousness. Nowadays they die straight off. Even young

people die off. The least thing kills them. In the old days even if a man were very badly wounded in battle he would recover. Men are mere weaklings (*amacoboka*) now. In these days . . . we do not see *amadhlozi*; we do not know where they have gone to; they left us with death. And we have no cattle to kill for them as, according to our beliefs, we ought to do.[120]

It was only during this period that Africans, en masse, began looking for alternative means of making rain, curing illness, and sniffing out "witches."

While some Africans converted to Christianity, others began to blend together Christian and African traditions. In 1904, Daniel Bryant of a Zionist sect from the United States baptized twenty-seven African converts. This sect, which practiced divination and baptism by total immersion in water, merged its beliefs with those of the American Pentecostals in 1908. The Zionist churches that now populate southern Africa emerged out of this second group.[121] An instrumental part of Zionism was its ability to exorcise African communities of witches. Missionary Bengt Sundkler remarked that in "general appearance, behavior and activities, the Zionists [were] . . . a modern movement of witch-finders."[122] The phenomenon of blending Christianity with witch finding can be seen from Mlenjeni among the Xhosa in the 1850s to the observations of anthropologists in the 1930s in central and eastern Africa.

Sundkler's work, *Bantu Prophets in South Africa*, draws distinct parallels between Zionism and traditional African cosmology. Sundkler, who worked as a missionary in Zululand from 1937 to 1942, found the initiation and healing practices of the Zionist prophet and the isangoma strikingly similar. Both focused on healing, though the Zionist prophet replaced the ancestors with the Holy Spirit.[123] One major difference, however, was that the isangomas of this time period were overwhelmingly women, whereas the prophets of the Zionist movement tended to be men. Zionism emerged at the turn of the century in the context of a growing number of separatist and independent African churches, growing hardships, and in the wake of multiple antiwitchcraft laws. Because it was Christian, Zionism thus became a legal and acceptable means by which Africans could sniff out umthakathis. Health was redeemed not with medicines or the killing or banishment of witches but by being beaten with holy sticks and partaking in purification rites.[124] Today's umthandazis, or faith healers, emerged from the Zionist churches and continue to represent another strategy for coping with illness. While some umthandazis use one or two herbal medicines particularly aloe, most rely only on prayer, holy water, and ash.

BIOMEDICAL CURES FOR WITCHCRAFT:
BIOMEDICINE AS TOOL OF EMPIRE?

By the late 1880s, colonial officials decided that antiwitchcraft laws aimed at reducing the powers of healers had had little effect in changing African behavior and belief systems with regard to witchcraft. The 1881 Natal Native Commission showed that many missionaries and colonial officials were still dealing with the problem of witchcraft accusations even though "smelling outs" had become more clandestine. Administrators thus proposed the introduction of biomedicine to the African community as a means of weaning Africans from African healers and blaming illness on witchcraft.[125] The idea of introducing biomedicine to the African population was thus motivated by three main concerns. The first involved replacing African healers with biomedical doctors in an effort to diminish the power of healers and consequently the power of hereditary chiefs. The second sought to combat African "superstition" through the introduction of biomedical doctors, who would, through their rational approach to the body, health, and wellness, discredit African healers. The third addressed the maintenance of African health, particularly as it affected the white population; this theme became more prominent during the twentieth century, when industries sought to create a healthier and more productive workforce. In 1881, the Natal Native Commission asked missionaries, government officials, and African chiefs to speculate on the effects of employing white doctors to discourage African belief in witchcraft.[126] This argument was not new but had been advanced as early as 1858 by Dr. Fitzgerald of the Eastern Cape who wrote that hospitals would "draw the savage from the remotest parts of South Africa and attach him forever to that Government which entered in spirit into his sickness and sufferings and provided a remedy." The answers of the 1881 commission encouraged this idea. Few questions asked by the commission actually addressed the health of Africans or the sanitary conditions of African homes and workplaces. This neglect can be contrasted to the Indian Immigrants Commission of the same year, which was very much interested in the availability of biomedicine and general sanitation of Indian immigrants.[127] The promotion of biomedicine by doctors and government administrators as a means of changing African cultural practices continued throughout the twentieth century.[128] While much of the rhetoric regarding the benefits of biomedicine focused on eliminating witchcraft, an increase of cross-cultural interactions in the urban areas and the emergence of an industrialized workforce brought a new interest in African health standards. This is not to suggest that the government developed an actual interest in the well-being of Africans; rather, their interest reflected growing concerns over Africans' role in the spread of communicable diseases to the white population and a desire to develop and maintain a healthier industrial workforce.[129] This trend can be seen in the early donations of mining money to the building of African hospitals in Johannesburg and contributions to the South African Institute for Medical Research in 1912.[130]

Lacking the will, money, and doctors necessary to saturate the African community with biomedical practitioners, the Natal legislature licensed those African healers who posed the least threat to the colonial state and most closely resembled biomedical practitioners. The 1891 Natal Native Code thus decriminalized and allowed for the licensing of a limited number of inyangas. There are no documents that address the thinking behind the decision to license inyangas, though it had been suggested previously by various magistrates. In 1881 a magistrate suggested licensing inyangas to limit their "mischief" and large numbers.[131] By limiting their type and number, colonial officials could ostensibly control the social and political power of African healers while maintaining a minimal degree of healthcare. After all, they reasoned, the medicinal practices of healers isolated from their political and social functions posed little threat to the colonial state. The licensing of healers was unique to the colony of Natal and Zululand, and these remained the only areas in a future South Africa where African healers were licensed and allowed to practice openly. The uniqueness of this legislation can be linked to both the powerful role of healers in the Zulu kingdom and the preceding Shepstonian system that advocated the use of African customary law as a means of governance. As customary law sanctioned the political power of those chiefs most disposed to the colonial state, licensing sanctioned those healers who were least threatening.

"TEA AND WITCHES": ENCOUNTERS OF CULTURAL CURIOSITY

At the same time that missionaries and government officials saw African healers as a potentially destabilizing force, the white public viewed healers as somewhat of a cultural curiosity. In many places around the world, dominant classes often come to romanticize subjugated indigenous cultures, particularly as they are seen to pose less and less of a threat. This seemed to be true for whites in Natal with regard to Africans in the colony, though the neighboring and independent Zulu kingdom was always held out as a potential threat.

Lady Barker's tea party, for which she invited the local isangomas to serve as entertainment for her white guests, reflects some of these changes in intercultural encounters, particularly white attitudes that mixed curiosity and disdain

Cultural curiosities: Pictures of healers often called "witchdoctors" captured the imagination of white settlers. B. Kisch, *Photos of Natal and Zululand* (1882). Killie Campbell Africana Library, C8102 (*above*); o C58–001–166 (*below, left*); C58–001–166 (*below, right*).

for African culture. By 1876, there was recognition by both the African participants and white observers that the "sniffing out" event was a cultural performance. Barker clarified this at the outset, when she told the Africans present,
"The only reason I had wanted to see them arose from pure curiosity to know
what they looked like, how they were dressed, and so forth. . . . This was only
a play and a pretence." The African performers readily agreed, and, given the
illicit nature of the performance and the prosecution of isangomas occurring
during this period, this was to a certain extent an absolute necessity. Barker's
insistence in telling the Africans, her guests, and her readers "that it was all
nonsense and very wrong" follows in the convention of many colonial memoirs in which authors attempted to distance themselves from local beliefs and
practices. Barker, like many whites, positions herself as a mere observer, un-
contaminated by the colonial experience.

As with many of the current tourist attractions in South Africa, there have
always been people willing to perform their culture for outsiders. In this case
the isangomas performed their culture because, in Barker's paraphrase of the
isangoma, "An English lady who loved our people wished to see and witness
our custom." Like the isangoma who chastised Isaacs forty years before, she
added, "White people do not believe in our powers, and think that we are
mad; but still we know it is not so, and that we really have the powers we profess."[132] While many Africans may have felt compelled to perform their culture—due to physical or monetary coercion—it also seems that others did so
out of a desire to celebrate a bygone era. With the rapid changes in Natal in
the nineteenth and twentieth centuries, a certain nostalgia may have emerged;
participation in such cultural performances validated African culture not
only to themselves but to an oppressive colonizer class. In this case, it seems
material benefits as well as an opportunity to vindicate their cultural practices
and beliefs may have provided the impetus to engage in such an event.

This spectacle of whites inviting, observing, testing, and at times sabotag-
ing the work of isangomas, as Isaacs had in the 1830s, appears again and again
in colonial writing.[133] The same fascination that brought Europeans—unable
to travel abroad—to the circus-like shows of London and Paris attracted Barker
and her guests to see the isangomas at her tea party. Barker's guests came to
gaze at the exotic Other, and, as Barker mused, her guests arrived "clam-
ourous to see the witches, wanting their fortunes told, and their lost trinkets
found."[134] Barker's use of the isangomas as spectacle—part entertainment and
part fascination—thus follows in a long tradition in which the dominant powers, by virtue of their position in power, had been able to subjugate and ob-
jectify the Other, through either public shows/exhibitions (beginning in the
early 1500s), photo postcards (beginning in the 1880s), or present-day cultural

parks such as Kagga Kamma in the Eastern Cape and Shakaland in KwaZulu-Natal (begun during the apartheid era).[135] All of these formats tried to capture what was different, exotic, and eccentric about the African "Other," and, like Barker's party, all were clearly staged for maximum effect.

Colonial-era shows combined fascination with fear. As Barker tells her reader, the knowledge that these women had at previous times caused other Africans "to sit trembling and in fear of their lives" created "a strong under-current of interest and excitement."[136] Likewise in 1853 eleven Africans from Port Natal exhibited "Caffre life" at St. George's Gallery in Hyde Park Corner, London. One of the acts included a "witch-finder . . . to discover the culprit whose magic has brought sickness into the tribe."[137] This act must have been particularly awe inspiring to a British public, as they were reminded of the contemporary resistance against the British led by a Xhosa prophet in the Cape Province. For Barker and her fellow compatriots in 1876, however, the isangomas exemplified the most exotic, at times terrifying, and now forbidden aspect of African life. On the inside cover of Barker's book, published in 1883, is a picture that could well have been the spectacle at her tea party event (absent the European tea drinkers). Clearly Barker, or her editor, knew this picture would appeal to British audiences in a post–Anglo-Zulu War period for the same reason that Barker's tea-and-witches party had appealed to her friends before the war.

The isangomas appealed to European sensibilities due to their "exotic" appearance, their transgression of normative gender roles—both Zulu and British—and their power in precolonial times to assert accusations that often resulted in the death or banishment of the accused. At a time when Africans in the colony of Natal had become familiar—wearing European dress, living and working in European homes—the isangomas remained outside this sphere and thus were distinctly different and less knowable. In describing their encounters with isangomas, Europeans often devoted a paragraph or more to describing the dress and appearance of these healers. Below are some fairly representative examples:

> Their clothing is hideous, consisting of skins and crocodiles and pythons.[138]

> Among this floating and thick plumage, small bladders were interspersed, and skewers or pins fashioned out of tusks.[139]

> Across her shoulders, and round her chest, she had pieces of twisted ox-hide, to which were attached snuff-boxes, bits of roots (medicine), bones, beads, and trash of various kinds.[140]

A glimpse of Lady Barker's "Tea and Witches" party? Found on the inside cover of
Lady Barker, A Year's Housekeeping in South Africa (1877; repr., London, 1883).

Given the powers exerted by these healers in precolonial times and in the
Zulu kingdom, as well as women's fairly limited role in public life, these
women's position in society posed an interesting transgression. When re-
marking on the airs of the isangomas, Barker states, "Their pride is to be
looked upon as men when once they take up this dread profession . . . They
are permitted to bear shield and spear as warriors, and hunt and kill with their
own hands the wild beasts and reptiles whose skins they wear."[141] In two societies,
Zulu and British, each based on rigid gender distinctions, the isangomas were
indeed curious.

The holding of Barker's tea party and her guests' reaction can also be interpreted as symptomatic of the colonial discourse and mind set of this period. As psychiatrist and Algerian revolutionary Frantz Fanon suggested, the
Manichean characteristics created by European colonial discourse during the
late nineteenth and twentieth centuries depended on defining European/white
as the antithesis of African/black. White symbolized everything that was pure,
rational, civilized, and good, while black symbolized savagery, irrationality,
and evil.[142] As Barker and her compatriots enjoyed a civilized cup of tea, all
their notions of "untamed Africa" were literally exhibited at their feet. Such
dichotomist notions engendered during the colonial era were central to ra-
tionalizing the existence and expansion of European colonialism. As Ann
Stoler shows in her work, ideas of European modernity and "Europeanness"
found their definition in the colonial experience.[143] By juxtaposing "Europe"
to "Africa," the British convinced themselves that British culture was more

desirable and civilized. Healers featured prominently in this discourse and were blamed for promoting and perpetuating African superstition. African resistance and obstinacy against white settlement, rule, and law were thus excused under the rubric of superstition and irrationality, rather than attributed to Africans' exercising their cultural free will or, in other cases, resisting white domination. Even in 1906, after the power of chiefs and healers had been largely curtailed, Jeff Guy demonstrates how white courts and administrators sought to blame a local uprising on the venerable "witchdoctor," rather than on hardships imposed by a newly implemented poll tax.[144]

Furthermore there were practical aspects of juxtaposing and emphasizing the incompatibility of these cultures. Colonists in the nineteenth century believed that cultural and bodily immersion in the tropics could result in illness. Colonists were thus rotated in and out of the colonies, and many sought refuge in the colonial hill stations, which provided cooler climes and a respite from tropical diseases and the stresses of colonial life. Such rotations and visits were also intended to prevent colonists from "going native," or sympathizing too

White settlers' drawings of the Other, fearful yet exotic. From Lady Barker, *A Year's Housekeeping in South Africa* (1877; repr., London, 1883).

much with the local population, and provided an opportunity to reinvigorate the spirit by retreating to a European or European-like atmosphere. In this way rotations or hill stations, like Barker's tea party, helped colonists to remember and reassert their own cultural and national identities. The necessity of distinguishing between the colonized and the colonizer made such movements and events a necessary part of the colonial experience. Lady Barker and her guests not only partook in a time-honored British tradition of teatime but were reminded of their own ideals and culture as they observed what they considered one of the most abhorrent aspects of the Other.[145]

CONCLUSIONS

Imagined as the epitome of African superstition, healers were often blamed for everything from the failure of missionaries to win converts to the inability of biomedical doctors to attract patients or of employers to retain laborers. Likewise superstition was used to justify increasing taxes, coercive labor, and the implementation of colonial rule. Healers were an easy target and initially provided missionaries, doctors, and employers an unproblematic explanation for their troubles. As the colonial government of the early twentieth century began to investigate the workings of African religion and healers and the indigenous judicial process, a more nuanced understanding of healers emerged though one that remained equally unsympathetic. Later, many anthropologists' insistence that African medical systems were based on superstition and witchcraft and provided only social and cultural remedies lent scientific credence to many of the stereotypes developed during the nineteenth century. Given the different discourses that have developed over the years, particularly among outsiders to African culture, it is little wonder that so many public-health initiatives sponsored by international nongovernmental agencies or by local biomedical programs have until recently approached Africans as if they have no viable medical systems of their own. This often resulted in culturally insensitive public health messages that were less than effective and the diversion of millions of dollars' worth of developmental assistance from indigenous healthcare providers.[146]

As these stories of early medical and cultural encounters demonstrate, Africans and Europeans were both important actors in determining and negotiating the tenor of their interactions. In the beginning we see that Europeans had little power: the female healer is able to literally negotiate for a better price from Isaacs, and he too has to settle the issue of witchcraft to maintain his legitimacy. He, like the colonial officials that followed him, essentially resolved this issue by recognizing witchcraft, but in the idiom of poisoning rather than witches. The further entrenched European power became, the less willing colonial officials were to negotiate. This shift in power led Africans to openly negotiate with Europeans, asking them to rethink their criminalization of healers and rain doctors—pleading, "It will help your trees too!" But it also encouraged subtler forms of negotiation such as the adoption of new medical diagnostics, as well as the examination of a new religion and new explanations for illness and misfortune. By the time of Lady Barker's "tea and witches" party in 1876, it would seem that Natal healers had largely lost their power. Yet this was clearly not the case given the continued concern over hereditary chiefs and the eventual licensing of inyangas in 1891.

From the period of initial contact and white chiefs in Natal, isangomas and witchcraft accusations constituted a political threat as they interfered with the implementation of white rule. The reasons behind colonial opposition to healers and local methods of eradicating umthakathis varied over time. Initial complaints focused on the injustice of killing alleged witches and later, under colonial rule, of chasing them out of the community or confiscating their cattle. These actions were regarded as odious by colonial administrators, although they sometimes participated in them (for instance, the case of Falibi). As colonialism sought to assert itself, Britain desired to strengthen its own rule by weakening the chiefs. While the British had developed customary law as a means of ruling the colony, it became apparent that they could not be effective colonists without legal recourse. Consequently they insisted that chiefs decide only civil cases, while criminal cases were taken up in courts overseen by white judges. Later, even this was taken away and chiefs were left with little more than symbolic power or whatever power they exercised surreptitiously. The criminalizing of healers and the decision to license inyangas were directly related to these political concerns. In response to this criminalization, in tandem with natural disasters and increased colonial encroachment on African lands, many Africans experimented with new forms of healing, either African, biomedical, Christian, or a mixture of all. As we shall see in the next chapter, the licensing of herbalists proved more complicated than colonial officials imagined. The resulting commercial and professional development of African inyangas in Natal brought them directly into commercial competition with white biomedical doctors and chemists in urban areas in the early part of the twentieth century.

4 ⌁ Competition, Race, and Professionalization

African Healers and White Medical Practitioners, 1891–1948

IN APRIL 1938, a Natal magistrate charged renowned African inyanga Bramwell Sikakane with twenty-one counts of practicing as a "native medicine man" or herbalist without a license. Although Mr. Sikakane had been found a competent inyanga by the Natal Native Medical Association in 1936, his application for a government license to practice as an herbalist had been repeatedly refused under the South African Medical, Dental and Pharmacy Act of 1928. During his trial, witnesses testified not only to Mr. Sikakane's extraordinary abilities to heal, but to the fact that he did what other African medical men in the district allegedly were unable to do: he diagnosed illnesses—often with the aid of a stethoscope and thermometer. Coming to testify in his defense, the executive committee of the Natal Native Medical Association arrived at court in "two 1938 model sedans—one a Lincoln Zephyr—and dressed in fashionably tailored European suits."[1] They came not only to lend support to one of their members, but to use this trial as a showcase to demonstrate against government interference in what they termed "native medical rights."[2]

The African healers described in this *Sunday Times* article deviated sharply from what the architects of the 1891 Natal Native Code had envisioned when they first licensed healers. Indeed by the 1930s the relationship between healers, government administrators and biomedical doctors had changed dramatically from the late nineteenth century. This was due in part to the creation in 1910 of South Africa as a larger union that combined the former British colonies of the Cape and Natal with the Boer territories of the Transvaal and Orange Free State, a union which then gained independence from Britain in 1910. The accompanying move from British colonialism to local white rule, followed by the professionalization of biomedicine, led local administrators to have less sympathy and say with regard to African inyangas. In this chapter, I investigate the development of competition between African and white medical practitioners and the role competition played in constructing local biomedical and African ideas of medical authority in Natal during the early twentieth century. I begin by describing some of the effects that licensing had on isangomas in the late-nineteenth century that enabled inyangas in Natal and Zululand to increase their status over other types of African healers in South Africa. Second, I examine the effects of urbanization and the unique government licensing of African healers on the commercial and professional development of African inyangas in this province. Third, I show how developments in African therapeutics threatened the commercial and ideological basis of white biomedicine. Fourth, I argue that this competition was vital to the development of local biomedical practices and African ideas of medical authority and contributed to the professionalization of both types of medical practitioners. Competition, racism, and the exclusion of Africans from biomedicine threatened African healers who nevertheless thrived by exploiting tensions that existed within the colonial administration. Fifth, I show how elements of gender, race, and class were employed in the construction of medical authority. Last, I illustrate that the process by which biomedicine became dominant in South Africa was not inevitable but was the result of intervention by South African physicians and pharmacists who worked to create and promote a unique form of medical authority that rested as much on racial difference as it did on rationality and science.

THE EFFECTS OF LICENSING INYANGAS AND URBANIZATION

As a result of the 1891 Natal Native Code, healers' roles and status drastically changed during the twentieth century. Whereas different types of healers had previously collaborated to maintain African health, new colonial laws forced inyangas to operate alone. The distinction between herbalists and African healers who used "supernatural" forces—isangomas and rain-doctors—enabled colonial officials to split the African healing community. Healers hoping to acquire government licenses and avoid legal prosecution began to adopt these terms and redefine themselves in relation to other types of healers. The term *isanuse*, which had been used interchangeably with the word *isangoma*, or *inyanga yokubula*, during the nineteenth century,[3] fell out of common use in favor of the term *isangoma* in the twentieth century. This indicated an important shift in the practices of these healers, from one of public "smelling out" ceremonies that detected umthakathis to private consultations to help mend relations with one's idlozis. With the 1891 legislation many isangomas tried to pass as government-defined inyangas. This legislative measure to weaken the power of the chieftaincy while maintaining a minimal degree of African health, however, became a permanent feature in Natal and Zululand that continued long after its initial implementation and affected the practice of both African and white medical practitioners.

After a long period of criminalization, one might expect that government recognition of any type of African healer would have been welcomed by African communities; this was not the case. Before the Natal Native Affairs Commission of 1906–7, set up to investigate the causes of the 1906 Maphumulo uprising, Africans complained that such licensing had done nothing to combat the problem of umthakathis but had merely tied the hands of the chiefs. "In former days these wicked people were put to death," but now "Chiefs, knowing the view of the Government in this matter, purposely refrain from saying anything about it."[4] In their continued efforts to convince the government of the irrationality of criminalizing witch-finders, exasperated witnesses argued "there are some diseases that are caused by natural means but most are brought about by human agency."[5] Magistrates not only refused to acknowledge the existence of witches, but "if one accused another of being an umthakati, the Magistrate ordered him to pay compensation."[6] Furthermore, the licensing of inyangas had caused undue hardship for both healers and patients. Mvuyana under Chief Kotshi asked, "How did it come about that a man is obliged to pay £3 for a license when he is simply in search of medicines for the purpose of doctoring his own people?"[7] "Why," asked Bewula of Umzinto, "is it necessary that the Government should impose this fee?" After all, "these men . . . exist for the general benefit of the Native public."[8] A larger problem developed, however, as the government did not seem to distinguish between people who used folk remedies and medical specialists, leading to the arrest of ordinary people.[9] This refusal "to allow Natives to doctor themselves by their own medicine"[10] seemed particularly ironic given that such government attentions to muthi did nothing to stop the dread of love charms. Instead, love charms seemed to have become more available, and men returning from the cities were using them on local women. This problem was further compounded when fathers attempting to bring charges before the magistrate were told that love charms, like umthakathis, did not exist.[11] Government's inaction against the rising tide of African ill health and death, combined with its dubious means of licensing inyangas, only increased African anxiety and concerns over colonial duplicity.

Legally, any African who gained the support of a local chief and magistrate could acquire an inyanga license. In practice, however, licenses were reserved for older men. Given that licensed inyangas often traveled and had ready access to cash, they had a large degree of independence. Chiefs and magistrates, struggling to maintain control of women and young men, limited licenses to older men. Government officials often rejected young men's applications on the grounds that they did not have enough healing experience, yet letters to and from the Public Health Department during the 1930s and '40s indicate that the real concern was a man's ability to contribute labor to the settler economy. When Tobias Mtshali applied for an inyanga's license he stated that he needed it as he could no longer perform manual labor. The magistrate denied his application after sending him for an examination by the district surgeon, who concluded that while he had a slight heart condition "there is no evidence of any lung trouble, and I see no reason that he should not be able to perform ordinary labor."[12] Women also faced a disadvantage since magistrates assumed that they were really isangomas posing as inyangas. While this may have been true, it left no option for the small percentage of female inyangas trained by their kin or husbands. These women often found themselves at the mercy of jealous husbands or male healers who had the ear of the male chief or magistrate.

The licensing of inyangas in Natal brought about a shift in government concerns. By the twentieth century, particularly after the failed uprising in Maphumulo in 1906, chiefs and the healers who supported them were no longer seen to pose a political threat to the colonial state. The political decline of the chiefs meant a similar decline in the prestige and power of the isangoma, isanuse, and inyanga yokubula. While such healers continued to unnerve missionaries, they became of little interest to the colonial government. Isangomas, who by the early twentieth century had become overwhelmingly female, continued to practice but did so within the private rather than public sphere. Occasional trials of healers over witchcraft accusations continued, but colonial officials found their attention increasingly focused on inyangas — licensed and not. While inyangas had been largely ignored by government officials during the nineteenth century, licensing brought up new issues that now mandated government attention in the twentieth century.

When Natal officials licensed healers they did not anticipate the many ways in which African inyangas would use the 1891 legislation to meet their own ends. Yet within a few years of this legislation, a district surgeon complained that licensed inyangas had begun leaving their kraals to "travel about the country looking for patients, promising cures, and saying the white man's medicine is no good for black people."[13] Prior to licensing, healers had remained home, venturing out to practice only when requested. Healers did not solicit patients,[14] but became known through work as an *uhlaka* (apprentice, or bag carrier) and through networks of friends and family. This changed during the 1890s. Droughts, locust infestations, rinderpest, and East Coast fever led people to welcome itinerant healers who touted new and unfamiliar muthi. South Africa's archival records are replete with complaints by Natal administrators regarding unknown traveling inyangas—African and Indian—offering their services to local chiefs and finding throngs of people willing to pay for muthi.[15] The rewards of this practice could be quite lucrative, and by 1915 it was African chiefs who complained of the lack of licensed inyangas within their districts.[16] Other itinerant healers, some licensed, some not, eventually made their way to the cities of Durban and Pietermaritzburg.

These two cities offered new sources of patronage and opportunities to Africans who migrated there; Durban, the larger of the two cities, tripled its African population from the 1880s to the 1920s.[17] After the devastating decade of the 1890s—drought, epizootics, and colonial encroachment on African lands—many Africans sought refuge and jobs in the cities. Initially attracted by wage labor, some men with specialized medical knowledge soon realized they could earn more money and enjoy greater independence by practicing as licensed inyangas. Other inyangas came to the city seeking new markets but particularly Africans with expendable income. The city proved a dangerous place. Not only were individuals afflicted by known maladies and new ailments caused by unhealthy urban living and working conditions, but unfamiliar people could be potentially carrying or administering dangerous muthi. Clients thus also needed medicines to protect them from unseen negative forces like witchcraft and love charms, as well as medicines that enabled them win jobs or avoid the police. While women healers may also have moved to the cities, late nineteenth and early twentieth century colonial policies generally discouraged the movement of women, who were deemed less useful to the urban workforce.

When African healers moved to the cities, they lost access to many of the herbs growing in the rural areas but often gained a wider knowledge base of available herbs. While healers continued to collect local muthi, they also bought exotic herbs from migrant workers or women visiting their families who came to town from the country.[18] Protus Cele, referring to his father's herbal gathering techniques explained, "He use to take a rope and tie himself—he had one hand—he'd tie himself here on his legs and tie on the tree and he use to go right down to the bird cliffs to collect all these *ipuchu*. He collected these things around by our home and in the area surrounding Durban."[19] The transportation of rural herbs quickly spawned a commercial trade in muthi. Persons gathering massive quantities of one or two types of herbs or animal fats could hawk their specialized items in the city for a large profit.[20]

Cities like Pietermaritzburg and Durban encouraged this trade by creating stalls and selling daily licenses for people to peddle muthi at the local Eating-Houses and Beer Halls.[21] Here persons could gain access to familiar folk remedies as well the advice of medicine sellers. The muthi trade increased the availability and variety of herbs coming into the city and enabled inyangas to expand their services and offer new curatives, much like the healers of the Zulu kingdom had expanded their knowledge and practice with the consolidation and expansion of the Zulu state. Yet there were certain dangers associated with losing contact with the collection process. Given that the strength of an herb is dependent on where and when it is grown and how it is collected, those who did not do the collecting possessed less awareness of the strength of an herb, which increased the likelihood of accidental poisoning.[22] Accidental poisonings may also have resulted from untrained persons posing as knowledgeable healers. While rural communities generally expelled useless or harmful healers, the large size and transient population of the towns and cities created less accountability.

The improvisations of the city not only involved new ways of purchasing herbs used in folk remedies and specialized medicines, but changed the ways in which some healers learned their profession. One such example involves Paul Cele, who started his own muthi business in 1938. Mr. Cele suffered

Muthi market. An inyanga and chief Ogle's son selling herbs in 1916. From Pietermaritzburg Archive Repository, American Board of Foreign Missions accession photo dated January 16, 1918, C.5583.

greatly as only two of his fourteen children survived past age six, and in desperation he approached an inyanga, Mr. Sibisi. After Sibisi apparently succeeded at controlling the illness and misfortune that had struck Cele's family, Cele asked him to show him how his medicine worked. His son, Protus Cele, describes how his father became a well-known herbalist in Durban:

> My father said [to Sibisi], if you die who is going to help me? You better teach me. Maybe I don't know who will die first, but if I die first you will look after my children, and if you die first then I will look after your children. Teach me the knowledge. Then Sibisi said, "I won't teach you, I'll give you a place so you can keep your table, lay your mat, display your plants, your medicine, and you will learn like that." And he started like that, people they come and buy and he asked people, what are you going to do with this and people would say I mix like this and then I do like this and this and it helps here. He took out a piece of paper and pen and write everything, then his knowledge was advance.[23]

Cele, who started selling herbs in stalls at Durban's eMatsheni market, eventually opened one of the first African muthi shops in Durban.

The commercial development of African medicine flourished as licensed African healers sought to distinguish themselves in the cities and smaller towns of Natal, particularly from the late 1920s and onwards. In the rural areas, people chose an African medical practitioner by his or her reputation. The anonymity of the city, however, forced inyangas to devise new means of winning the trust of urban folk. In the breach, some healers turned to advertising in leaflets and in the African press. In the 1930s they began to open up shops in downtown Durban, listing themselves in the local directory as "native chemists."[24] Many also erected signboards outside their shops that noted their expertise and medical license. African healers in the city also attracted customers by continuing to incorporate new forms of healing they learned from other isangomas and inyangas and from white chemists and doctors. In particular, they appealed to modernity by borrowing the implements and language of biomedicine and science. Some inyangas bottled herbs and used preservatives, stethoscopes, thermometers, and other modern equipment.[25] Ngcobo, for instance, had advertised himself as a "doctor of native medical science."[26]

Healers worked hard to carve out this new commercial niche. They sought to appeal to a more sophisticated urban African community while legitimating their trade with age-old notions of tradition. Israel Alexander, a successful

Sotho healer who changed his name and opened up several herbal shops in Durban in the 1930s and '40s, ran a successful mail-order business. In 1932 he even listed himself in the city directory under "European Chemists," though the error was caught and corrected by the next year. One of Alexander's pamphlets displayed symbols of biomedicine—bottles, pills, and biomedical diagrams of internal organs—with pictures of a "traditionally" dressed Zulu maiden fetching water and an African chief outfitted in a traditional leopard skin and feathered headwear. He appealed to urbanites with complexion creams and hair pomade and showed Mr. Liphapang—a pleased customer— clad in a dapper white suit and sporting a fashionable hat. But he also tapped into public notions regarding the value of traditional medicines by emphasizing that "All remedies are guaranteed pure and free from harmful drugs."[27] In essence, Alexander attempted to bridge the symbols and perceived advantages of modernity with the professed values and stability of African tradition.

Urbanization, and the competition that accompanied it, thus required commodification, commercialization, and changes in therapeutic techniques. With so many African and white medical practitioners vying for clients and making fantastic claims, city dwellers—African, white, and Indian—needed a means to distinguish between persons who just wanted their money and the more legitimate and efficacious healers. Many who used African urban medical practitioners came to rely on government-issued licenses, most likely because licensed inyangas themselves presented them as symbols of legitimacy and efficacy.

COMPETITION BETWEEN WHITE AND AFRICAN MEDICAL PRACTITIONERS

The licensing of inyangas in Natal and Zululand gave the inyangas (and those who moved there) a great advantage over isangomas and their counterparts in the rest of South Africa, but the ensuing transformation of African medicine also constituted a genuine commercial threat to the white biomedical community. This threat manifested itself in three major ways: first, white doctors perceived the loss of a traditional client base; second, competition for African clientele emerged between African healers, white doctors, and pharmacists; and third, African licensing and commercial practices challenged the scientific and moral "authority" of white biomedicine.

While the historiography of medicine in South Africa has acknowledged the competitive climate of the white South African biomedical community, competition between biomedical practitioners and Africans healers has been greatly understated.[28] Due to overcrowding of the medical field after the Anglo-Boer War (1899–1902), white biomedical practitioners found themselves quite sensitive to the presence of new doctors from overseas.[29] As I. S. Monamodi explores in his dissertation, such dire economic circumstances led doctors to divide themselves along ethnic lines—English, German, Dutch, and Jewish. After World War I and during the Depression this competition increased as doctors from overseas and graduates of the first medical school entered the medical market; this coincided with a decrease in demand and an impoverished client base that returned to home remedies. Such competitive conditions in the first part of the twentieth century among white medical practitioners caused great sensitivity to any perceived intrusions on their client base.

Within this plural medical society, inyangas were also aware that biomedicine could undermine their own status and economic advantages. Many Africans had begun to use European medicines when they were first introduced by white traders and missionaries, and later by district surgeons after the 1850s and '60s. Although many Africans remained skeptical and resisted the imposition of biomedicine via vaccines and quarantines, the Natal Native Blue Books show an increasing acceptance in certain areas from the 1870s to 1910. Vaccination, which remained voluntary for the white community, was enforced with the help of the police in African ones. Yet greater compliance emerged as Africans witnessed the success of the smallpox vaccine when the disease broke out in rural communities in the 1890s.[30] In the twentieth century, the availability of biomedicine to the African community also increased, through the building of both African hospitals in towns and cities and clinics in the rural areas. Furthermore, fear of communicable diseases led to the government's launching of a number of public health campaigns that sought to convince Africans of the efficacy of biomedicine as well as the importance of hygiene, vaccinations, and safe sexual practices.[31] More exposure, however, did not always mean an acceptance of biomedical concepts or practices.

As a means of countering competition with biomedicine, some healers discouraged Africans from attending white doctors. Katie Makanya, who worked with Dr. James McCord at his Durban dispensary, recalled in the first decades of the twentieth century that "at first all the inyangas claimed the Doctor knew only the white man's diseases and nothing of African sickness. They warned people not to go to him. Consequently, too many patients came only as a last resort."[32] Likewise a president of a local farmers' association complained that "native doctors" had kept Africans from using quinine during a malaria epidemic in Zululand in the late 1920s.[33] And in 1940, Edward Jali, one of the first African biomedically trained "medical assistants" complained in a letter to McCord's wife that he was refused permission to treat an African patient on the grounds that only an inyanga could cure such sickness.[34]

Other healers attempted to learn biomedical practices and improve their own therapies. Healer Thlambesine Ngcobo eagerly visited Dr. McCord's dispensary after hearing about it from his sister. The two medical practitioners showed each other remedies, after which Thlambesine Ngcobo began to refer patients to McCord.[35] Likewise Dr. Sutton complained of a referral from a "Zulu 'Dr.' [who] practically came along to consult with me regarding the patient."[36] This cooperative referral, however, seemed to go only one way—toward biomedicine. Other healers hired medical assistants who had previously worked for biomedical doctors. McCord's assistant Mpandlana was wooed by an inyanga who offered him fifteen pounds a month and, Makanya suspected, had the intention "to steal the Doctor's prescriptions and do what he's not allowed to."[37] Likewise Mafavuke Ngcobo himself had hired two assistants who had previously worked for chemists. Most importantly was Mizriam Mngadi, a former McCord assistant who began working with Mafavuke Ngcobo in 1930, after which Ngcobo reputedly expanded his remedies and began using chemist supplies bought from a variety of registered chemist supply stores.[38] One of Dr. Campbell's assistants opened his own practice based on Campbell's reputation among Africans,[39] and J. Gobezi, who came to work for Dr. McCord during the first decade of the twentieth century, not only learned how to mix prescriptions but began to sell biomedicines on the side. After being fired by McCord, Gobezi obtained an inyanga license and led patients to believe he also had the equivalent qualifications of a biomedical doctor.[40] Government chemists discovered that other healers such as Israel Alexander sold cough medicine that contained alcohol and quinine.[41] By combining the elements of traditional healing and biomedicine, these healers aimed to appeal to the rise of new and complicated needs and identities of African urban dwellers.

The licensing of inyangas and advertising prowess of African healers as well as their new visibility in the cities in the early twentieth century increased white anxieties as whites were also seen using African healers. Some whites had always visited African healers, particularly during the early years of contact and in outlying rural areas where biomedical doctors were few or nonexistent. Given the option, however, white doctors and government officials assumed whites (by virtue of their "race" and culture) would choose to attend a white medical practitioner. White patients visiting African healers led many white doctors to complain that inyangas had intruded on their traditional client base. Government officials, equally concerned, noted in 1911: "There are grounds for believing that in some cases in Natal white people of the poorer classes are being treated by these native doctors."[42]

African inyangas appealed to white clients of various ethnic backgrounds in a variety of ways. First, one of the main advantages in visiting an African healer was the lower cost. Second, home remedies of many poor whites resembled African medicine more closely than they resembled the remedies of the specialized biomedical physician of the early twentieth century.[43] Third, in the rural areas it was often easier to gain access to an inyanga than a district surgeon or white medical practitioner. Forth, after the Anglo-Boer War many Afrikaners, who prided themselves on self-reliance, may have been wary of biomedical doctors, who tended to be English-speakers. Fifth, inyangas often offered confidential services not easily obtained from biomedical practitioners—including herbs to counter sexual dysfunctions, offer mystical protection, and attract members of the opposite sex. For example, men used *bangalala* for greater sexual excitement,[44] *ndiyandiya* was recommended for those appearing before court and was said to "confuse a judicial officer and to sway him to their favour,"[45] and "Love Drops" were bought to attract the opposite sex.[46] While some whites openly espoused the use of African healers—going so far as to testify on the behalf of inyangas in court challenges—others preferred to secure muthi more discreetly, sending African workers to make inquiries on their behalf. Other whites obtained herbal remedies from the mail-order houses of Israel Alexander and Mafavuke Ngcobo, which began selling herbs through the post in the 1930s.

The medical treatment of whites by African inyangas drew criticism from the Natal Medical Council as early as 1905, when Mafayifa Radebe, an African healer from the rural district of Utrecht, sued Johannes Van der Merwe in a magisterial court for refusing to pay for medical services that resulted in Van der Merwe's recovery. While the magistrate initially dismissed the case, arguing that Radebe's license authorized him to practice only among Africans, the Supreme Court ordered a reappraisal, and Radebe won a settlement for £25.[47] Dr. Campbell Watt, head of the Natal Medical Council, wrote to the colonial secretary concerning this case: "The Medical Council views with disfavor the fact that Native Medicine men are placed on an equality with European Medical practitioners as regards suing for fees for attending European patients and considers that their license should be amended so as to indicate that their practice is restricted to Native patients."[48] The state seemed to respond to this request, as the number of inyanga licenses issued, which had increased every year since their introduction in 1891, dropped from 1,925 in 1905 to 1,496 in 1906.[49] Yet in 1909, when Europeans again gave depositions testifying to the healing abilities of Durban *"inyanga"* J. Gobozi, the Medical Council demanded an outright end to inyangas' licensing.[50] Gobozi, who had been in the employ of McCord, was said to practice along largely biomedical lines.[51] Although the number of white patients treated by African healers probably remained small, a crowded medical market meant any infringement was cause for complaint. Furthermore, insult was added to injury as the government sanctioned this unwanted intrusion.

As the number of white biomedical doctors and chemists in South Africa increased during the early twentieth century, they sought to expand biomedical markets and services. The most obvious solution was to include African clientele. White chemists began soliciting African clients through African language pamphlets and by placing advertisements in African newspapers in the 1910s and early 1920s. Others tried to persuade the literate classes through the African press that healers "kill/injure your wives through their folly and ignorance."[52] The local medical council, which preferred to think of physicians as altruistic rather than commercial, forbade doctors to advertise their services. White doctors nevertheless continued actively to seek African clients, and in one case this led two white doctors to write to the Natal Medical Council to resolve a dispute. In 1911, Dr. Bonfa accused Dr. Chambers of stealing patients by hiring a "Coloured" and an Indian who guided African patients to Chambers's office, thereby decreasing the number of available African patients.[53]

Much as early biomedical practitioners of the nineteenth century had sought to meet African expectations of proper medical treatment, biomedical practitioners of the twentieth century, such as McCord, conceded to "African psychology" and changed their practices and medicines. "In visiting heathen kraals I often had to diagnose without asking questions, simply because the witch doctor did and the Zulus were accustomed to the method."[54] McCord thus attempted to gathered information about his patients prior to seeing them so he could tell a story of their illness. As a missionary doctor McCord had considered offering free medical treatment, but, understanding that African ideas of medical authority regarded "free treatment or medicine as worthless," he decided to charge for his services and medicines.[55] During the malaria outbreaks of 1906 and 1930 McCord noted that Zulus "threw away the [government's] free tablets and came to the dispensary to buy my 'blue medicine' at twenty-five cents an ounce." He attributed this to two factors: he prepared quinine in a liquid form as Zulus "were accustomed to drinking medicine," and he added methyline blue aniline dye. While this dye had no medicinal properties, it had the visible effect of turning patients' urine blue. This, combined with the terrible taste, convinced African patients that this was strong medicine, which resulted in greater patient compliance. As McCord concludes, "They had no faith in either pills or free treatment, and quinine pills looked to them like a feeble remedy for a powerful sickness."[56] In another appeal to gain African clients, Durban chemist C. Melody was said to use African names in order to sell some of his (biomedical) medicines.[57] As late as the 1950s, H. C. Lugg, welfare officer at King Edward Hospital, told a patient still concerned about a witchcraft spell that the hospital "used a form of lightning for disposing of such spells." This lightening was in the form of shock therapy, which Lugg contended proved successful in eradicating his patient's fear.[58]

While local African medical cultures were fairly open and accepted certain aspects of biomedicine, many doctors and chemists found their efforts to break into the African market frustrated by African skepticism of biomedicine. Most Africans sought treatment from white doctors only for "white" diseases such as syphilis or when their own remedies had failed. Africans often perceived biomedical practices as ineffective and culturally alien (i.e., the consultation procedure or confinement), and even as the cause of numerous African deaths.[59] Hearing that the biomedical community intended to quarantine leprosy sufferers, Ndwedwe's magistrate reported in 1903 and 1905 that doctors were unable to find victims of the disease given that "unfortunate sufferers are carried daily at dawn, and hidden in bushes and rocks, being taken back to the huts after dark."[60] Chiefs, responsible for African well-being, thwarted vaccination attempts by encouraging noncompliance with biomedical authorities at late as 1929.[61] Until the 1930s and '40s, most Africans refused hospitalization, claiming that hospitals were houses of death.[62] And even in the mid-1940s, workers at the National Health Centre at Polela had trouble convincing parents that their children had not been sold but confined for T.B. treatment.[63] Wariness of biomedicine seemed to vary from place to place depending on biomedicine's success or failure with previous epidemic diseases like malaria and smallpox. Yet different ideas of the body and etiologies of illness and compulsory biomedical campaigns enforced by the police or accompanied by demands for payment for mandatory vaccinations had left many Africans resentful. Rather than blaming African incredulity on the coercive nature in which biomedicine was often introduced to the African community, Dr. Campbell Watt attributed it to the licensing of inyangas: "The recognition of this class of ignorant uncivilized medicine men and their system retards the progress in civilization of the native population generally, confirms their savage beliefs and prevents them from attaining to a proper appreciation of the advantages of modern medicine and surgery."[64] Licensing was seen as sanctioning African superstition and healers and as an obstacle that prevented white medical practitioners' access to the African market. African skepticism of biomedicine only increased as Durban imposed mandatory examinations on African men coming into the city after 1917 and began mandatory "dipping" or delousing of African bodies in 1923, something which previously had been reserved for livestock.[65] This final action caught the attention of the Industrial Commercial Union (ICU), the largest African labor and political movement at the time, which responded by passing a resolution requesting the immediate abolition of this "unreasonable, ignominious, inhuman and degrading action."[66]

By the 1920s not only were inyangas perceived as a commercial threat by a number of biomedical practitioners, but so were the licensing and commercial development that had lent inyangas medical legitimacy in the eyes of both African and white clients. Biomedically trained aides like E. Jali discovered that chiefs held licensed inyangas in higher esteem than Africans like himself. Indeed licensing seems to have reached its height in the early 1920s, with licensed inyangas numbering 2,137 per annum in 1921.[67] This supposed authority upset the rhetoric of biomedical practitioners who made their claims to efficacy and morality by defining themselves in contrast to African healers. The professionalization of biomedicine in South Africa needed a foil against which to project itself as "scientific," technologically advanced, and the sole possessor of knowledge on the body, health, and wellness. McCord's nurses even caricatured "witchdoctors" in weekly hospital plays, and Jali sought to convince local chiefs to use biomedicine by citing the ineffectiveness of inyangas.[68] While the biomedical profession defined itself in contradistinction to all nonallopathic practitioners or white "quacks," osteopaths, cancer curers, and Christian Scientists, they were derided for knowing better, whereas African healers were described as innately superstitious. White medical practitioners, eager to establish their own authority and expertise, encouraged a discourse that painted indigenous healers as malicious, unenlightened, and avaricious bunglers who manipulated a gullible public. This argument proved easier if African medical practitioners were thought of as "witchdoctors." But, by the 1920s, the distinction that white doctors sought to draw between "science" and "superstition," between "reason" and "irrationality," had become muddled as "witchdoctors" began to take on many of the outward appearances of the biomedical profession.

Dr. Gordon, the district surgeon of Krantzkop, complained in 1924 to the Natal Medical Council of such a healer: Ndlovu Dlamini, a licensed inyanga who displayed a huge signboard over his residence proclaiming in Zulu and English that he was a "licensed medical practitioner" skilled in "medical electricity." Though Dr. Gordon surely saw this "doctor's" use of a battery and compass for "electro-diagnostic" purposes as mere "quackery," his larger problem was that the public was unable to discern between "legitimate" and "illegitimate" medical practitioners. Dr. Gordon ended his letter: "It is surely disgraceful if an unqualified native is allowed to compete with a fully qualified European Doctor!"[69] While white medical practitioners chided African healers for being superstitious and perpetuating African ideas of witchcraft, they much preferred this form of "authenticity" to the syncretism occurring in the towns and cities where healers blended African therapeutics with African ideas of modernity and biomedicine.

Consequently, African healers found themselves accused by the biomedical community of not practicing "traditional" African medicine, but of adopting "biomedical practices" such as performing a "diagnosis." Prior to the licensing of inyangas, herbalists were generally consulted only after a diagnosis by oneself, one's family, or an isangoma.[70] Since colonial law had criminalized the isangoma (the diagnoser of African medicine), licensed inyangas were forced to learn the art of consultation. While many inyangas borrowed the techniques of the isangoma, such as "throwing the bones," others learned or appropriated those of biomedical practitioners, such as the stethoscope. Interestingly, both throwing the bones and the stethoscope were known as *inhlola*, instruments for ascertaining the hidden truth. By offering a diagnosis and using biomedical diagnostic tools, inyangas attracted the hostility of white doctors who claimed exclusive rights to "diagnosis" and "rational therapy."[71] Throughout the 1920s, physicians had challenged white chemists and patent medicine dealers in court for "diagnosing" or "practising as a doctor." Physicians argued that inyangas should likewise be limited to dispensing medicines. Colonial administrators, who favored the licensing of inyangas, did not agree, and in 1941 and 1943 magistrates ruled that an African herbalist was both a medical practitioner and a chemist rolled into one and thus had a legal right to diagnose.[72]

Israel Alexander's Shop at 920 Umgeni Road, Durban, 1931. South African National Archives Repository, NTS, 1/376, file 5.

Alexander's home in Sydenham, a suburb of Durban, 1940. Photo by Lynn Accutt. From *Natal Regional Survey: The Durban Housing Survey* (Durban: University of Natal Press, 1952), 320.

Pharmacists also lodged complaints about African healers, particularly as urban inyangas added chemicals and patent medicines to their businesses. In 1931, the Natal Pharmaceutical Society complained about two African healers who, advertising themselves as "qualified doctors," used and sold biomedical medicines in "shops with all the outward appearance of European chemist shops." The society claimed not to be "opposed to bona fide native doctors carrying on their practice in a bona fide native way," but was upset that these persons were "allowed to masquerade as fully competent medical practitioners and chemists on a European standard."[73] In 1938 the pharmaceutical society targeted five major African herbalist shops that sold patent medicines and mixed herbs with preservatives: "It is alarming to watch these herbalists steadily encroaching on our business and gaining ground, and it is time every chemist in Natal kept them at arm's length and made it impossible for them to obtain supplies."[74] The crowded medical market, as well as the increased visibility of urban African healers practicing syncretic medicine in the early twentieth century, led a number of white medical practitioners to conclude that African healers posed a genuine commercial threat that needed to be contained.

ESTABLISHING WHITE BIOMEDICAL AUTHORITY

Biomedicine not only sought to eliminate competition from African healers, but also desired to restrict medical authority to whites. This desire became apparent in the late 1920s as physicians debated the training of African doctors, orderlies, and nurses. While physicians agreed in principle that African and white doctors should meet the same educational standards, concern over competition, particularly from rural biomedical doctors, ensured that such equality would never occur in practice. Rather than competing against one another, white doctors began to turn in racial solidarity against all types of African medical practitioners. In 1913 the president of the Natal Pharmaceutical Society stated, "The pharmaceutics profession . . . had no option but to preserve it[self] for the white man."[75] Though a handful of African doctors trained in Europe and America gained recognition from the Medical Council, African South Africans found themselves excluded from South Africa's own medical schools until 1941 and from pharmaceutical apprenticeships required by all licensed chemists. In the debate over training African practitioners in biomedicine, a compromise resolved that locally biomedically trained Africans would service the African population through government employment. Though they had the educational and experiential equivalent of white doctors, they would not carry the title "doctor" but instead would be termed "medical aides," they could practice only under the supervision of a white government doctor, and they could not augment their incomes through private practice.[76]

The number of biomedically trained African doctors in the early twentieth century was negligible. Animosity toward them, however, stemmed from white fears that their numbers would increase as well as from the ideological and racial threat they potentially posed. These fears reflected those of a racially apprehensive society. White South Africans working to implement segregation and enforce their racial superiority were threatened by any transgression of racial and cultural barriers. Many whites assumed that Africans trained as biomedical doctors would become "deracialized," meaning they would become alienated from their own cultures, move to the cities, and acquire the tastes and desires of white South Africans.[77] When debating the merits of medical training for Africans overseas, Dr. James McCord argued, "There is a very distinct danger that five or six or seven years in Gr. Britain would get a man out of touch with his own people and lead him to acquire the ideas and tastes of a white man. He might even want to drive his own motor car and have his office in the Britannia Buildings. He would very likely be unfitted for the simple life desirable for the doctor depending on a family practice among the Zulus."[78] Inevitably, it was expected, black doctors would ignore racial boundaries, not only in their treatment of white patients, but in social and political spheres as well.

The Mafeking nursing strike of 1927, which witnessed white nurses protesting against the presence of African physician Dr. Molema at the Mafeking public hospital, brought these anxieties to national attention. The case of Dr. Molema was provocative; it is also instructive in what it tells us about white attitudes toward interracial interaction, especially when circumstances involved a black male professional. Biomedical doctors had worked hard to imbue their profession with a sense of authority and privilege within both the white and black communities. Members of the white power structure found it troubling that a black man, be he a doctor in a public hospital or an inyanga working in a "chemist" shop, could solicit that same respect and authority from patients or clients across the color line. Furthermore, as the case of Dr. Molema highlighted, they feared the impact of black doctors exercising this authority over white patients and white nurses. As Dr. Hay-Michel stated, "The point we want to explore is the effect upon the native mind of the privileges, responsibilities, and aspirations that men of their own race and colour acquire by entering such a profession, on terms of equality, or even in competition with Europeans, and how this would affect their attitude towards the dominant White population of this country."[79] Doctors and government officials alike feared the effect this might have on the social order.

To safeguard their profession in an increasingly polarized society, the biomedical community turned increasingly to the language of "racial science" to convince the government and others of their own unique medical authority. For instance, biomedically trained African doctors might be sufficient for Africans, but their alleged lack of moral authority meant they could never adequately care for white patients. As one physician stated, "There is no reason to suppose that the cultivation alone of the intellectual faculties of the native . . . will fit him for the social, moral and cultural requirements that medical practice among Europeans demands."[80] Because Africans were deemed innately superstitious they were automatically morally and scientifically suspect. As Dr. Gale argued in 1938 against the opening of an African medical school, African biomedical doctors would not be able to resist the temptation to supplement their income "by a judicious blending of orthodoxy with izinyanga-ism."[81] As Monamodi rightly points out, this was a way for white doctors to claim that they, by virtue of their cultural superiority, should have access to all population groups.[82]

Doctors also used the rhetoric of race to attack white patients who saw African practitioners. European medical professionals claimed that whites who visited African practitioners suffered from "racial degeneracy"; whites did not visit African doctors and healers because they were efficacious, inexpensive, or convenient, but because they had presumably adopted the "superstitions" of Africans and "Coloureds." In short, they had "gone native." Evidence that cultural exchange was occurring alarmed both the medical community and government officials. In an article submitted to *The Journal of the Medical Association of South Africa*, J. A. du Toit wrote:

It is just as impossible for the native to resist the temptation to treat a European as it is for some of our unfortunate and ignorant Europeans not to go to the native for treatment. The native doctor has attractions for the ignorant European by virtue of there being something mysterious about him, something of the "toordokter" [witchdoctor]. Hence the native will always get a small following of Europeans to stimulate his ambitions to go further, and if allowed to go on unhampered but encouraged by "his White colleagues" we Europeans will soon be the exploited and our hopes of a pure White race will soon be gone.[83]

For du Toit, white patients visiting black doctors would result in race suicide.

Alarm in the white medical community became particularly acute when racial mixing involved black men and white women. An inyanga, "Dr. Mpanga," raised the ire of the settler medical community not only by using the biomedical title "doctor" and a stethoscope, but by "attending European women in confinement, and apparently also being consulted by European women because of sterility!"[84] White liberals who responded to the Mafeking incident argued that "men like Dr. Molema would be wise if they recognized the prejudices that exist, and personally declined to treat European women in state-aided hospitals."[85] Again du Toit put it more bluntly: "We have a right to protect our race, and how can we when we allow our wives and daughters to be attended by native doctors?"[86] These sorts of arguments raised in the medical journals and in letters to government officials provided compelling reasons for government interference and protective legislation.

Interestingly, female African nurses tending to white men did not arouse the same indignation as racial mixing between black male medical practitioners and white female patients. Nurses were women, and played a subordinate role to doctors and administrators. In addition, their position as caretaker was familiar: wet nurse, nanny, and household domestic. Whites may have seen African women as sexual, but they did not pose a sexual threat, at least not in the hospital. African men, on the other hand, were scapegoated for social ills during the "Black perils" of the late nineteenth and early twentieth centuries in South Africa and seen by many in the white community as sexual predators. Writing of the Molema affair, historian Shula Marks states, "The status of black male doctors contradicted their racial position of inferiority; clearly white nurses found this intolerable. Moreover the vision . . . of black male hands on white female bodies, was powerfully disruptive of white notions of social order."[87] Advocates of biomedicine therefore appealed to whites' fears of racial degeneracy by citing African doctors' and healers' attendance to white women and the real or imagined sexual, social, and political implications of such interactions. Consequently, hapless white women featured prominently in the discourse surrounding biomedicine's medical authority and the limits placed on African inyangas and African biomedical practitioners in South Africa.

POLICING THE BOUNDARIES OF WHITE BIOMEDICAL AUTHORITY

Fearful that the color line dividing science from superstition, rationality from irrationality, medicine from quackery, and white from black was crumbling, the South African biomedical community sought to reinforce its own ideas of medical authority. Like their contemporaries in the U.S., they turned to the law to impose their standards.[88] Forming themselves into a national medical organization in 1926, the white medical community lobbied to ban African physicians from white public hospitals and pushed for an end to government licensing of African herbalists. Strong lobbying efforts by the newly formed South African Medical Association finally resulted in the passage of The Medical Dental and Pharmacy Act of 1928, which eliminated all types of medical practitioners not acknowledged by the association and clearly banned all variants of African healers throughout the union. Only currently licensed inyangas in Natal were exempted. The law, however, aimed to reduce the number of inyangas by revoking licenses not renewed within three months and requiring new applicants to apply through the minister of public health.

Like many healers, inyanga Mantsholo Dube, with the help of a solicitor, petitioned the Native Commissioner in 1937 to reinstate his license despite his failure to renew it on time. His letter and the refusal that followed were very typical of the letters that crossed the desk of the minister of public health. Dube appealed,

> Shortly after the Anglo-Boer War I was residing in the Dundee District, Natal, where I was a Licenced Native Medicine Man, Inyanga and carried on my business for several years . . . I then moved to the Utrecht District, Natal and there I also carried on business as a licenced Native medicine man Inyanga. I then moved into the Nqutu District and there I again carried on business . . . for 16 years paying for my licence each year. During January 1936 I became very ill and I was unable to travel and was prevented from applying for the Renewal of my Licence by illness. When I had partially recovered I went to the Native Commissioner's Office at Nqutu on the 11th day of May 1936 and applied for the Renewal of my said Licence but my

application was refused as per endorsement on my Licence for the year 1935. Owing to the refusal for the Renewal of my said Licence I am now deprived from carrying on my business as a Native Medical Man and am unable to make a living . . . The position now being that I am deprived of my living and have a Wife and family to support and am now unable to provide for them.[89]

Though Dube had been licensed for sixteen-plus years, he most likely also represented the type of healer the government sought to discourage—a somewhat transient inyanga and one who we can infer from his letter was most likely of working age. Unlike the Natal and Zululand chiefs and magistrates who previously issued inyanga licenses and showed sympathy for African cultural practices and local medical needs, the minister of public health sympathized with the biomedical community. He aimed to eliminate competition between white doctors and African healers. The government issued few new licenses, and, when it did, the process usually took several months and required that the inyanga hire a solicitor. Licenses were granted mainly in rural areas where few biomedical doctors practiced. As a result of the 1928 law, the number of yearly licensed inyangas dropped from 1,000 in 1928 to 566 in 1932.[90] In Durban only 4 licensed inyangas managed to survive in 1940 to serve a city with 66,993 Africans.[91] Consequently, African healers found new ways to sell their wares, some illegally and some through the acquisition of general dealer's licenses.

Durban chemists spearheaded further legislation in the 1930s that successfully outlawed the use of "European medicines" by licensed inyangas. What constituted "European medicines" is another topic unto itself, as ideas of medical borrowing did not flow just one way. Inyangas were also prohibited from using the title "doctor" and forced to limit their practice to African patients.[92] The biomedical community also lobbied, less successfully, to restrict the practice of inyangas to the district in which their licenses were issued,[93] to ban licensed herbalists from urban areas,[94] and to prevent African use of stethoscopes and bottles.[95]

The Pharmaceutical Society of Durban decided in 1930 to deter the proliferation of licensed herbalists by establishing a "Native and European Herbalist Committee." Members of this committee searched the newspapers for all licensing applications that included the selling of herbs or muthi and then lobbied local government to oppose such licenses. They also conducted their own examinations of African herbalist shops and demanded to be shown licenses. Africans without a license and those selling "European medicines" (such as Vicks VapoRub, castor oil, or anything with alcohol or preservatives) were promptly reported to the Criminal Investigations Department, Department of Public Health, and/or the Pharmacy Board.[96] In December 1939, the president of the society commended the herbalist committee: "We have been interested in the elimination of the so-called native medicine man and, while we have not been successful in closing them up, we have most certainly curtailed their scope to a great extent, with, of course, the assistance of the Pharmacy Board and Department of Public Health.[97] The following year the society noted with concern that the attorney general had hesitated to proceed on the case of Mafavuke Ngcobo, at which point they agreed not only to help provide a chemical analysis of the mixtures in Ngcobo's shop, but to pay any related expenses![98] Although Ngcobo's conviction, as discussed earlier, succeeded in legally restricting "traditional" medicines, the number of muthi shops greatly increased during the 1940s, declining only with the implementation of the Group Areas Act in 1950.

White physicians and chemists claimed the new laws and precautions were needed to protect the public good, but many African healers saw a more sinister motive at work. Licensed inyangas such as Mafavuke Ngcobo, who read the professional medical journals, knew that the new laws reflected an attempt by the white medical establishment to stamp out "commercial rivalry." In a letter to the minister of native affairs, he complained, "It is unnecessary to look further than certain of the European Chemists of Durban and Pietermaritzburg for the source of the agitation. For some years past it has been the aim of these Chemists to win the Native population to the use of European remedies, and for this purpose they have engaged a large staff of educated Natives, who are constantly employed in circularizing the Natives of South Africa and in replying to correspondence arising therefrom."[99] Ngcobo's analysis of the situation, however, did not go far enough. Not only did white medical practitioners want to convert the African population to the use of European remedies, they also wanted to cash in on the muthi trade. It is ironic that the pharmaceutical society pursued legislation and harassed herbalists at the same moment that certain white chemists, including a former president of the society, were busy selling muthi to Africans through the post! W. R. Pimm & Co.'s Kwa Ndhlulamiti pamphlet, written in 1920 in Zulu, for instance, advertised products such as the fat of wild animals and tokoloshe.[100]

Interestingly, these pamphlets from white pharmacists violated all the moral and scientific rationales that biomedicine claimed distinguished it from African therapeutics. For instance, they acknowledged and purported to cure local culture-bound illnesses such as idliso or the effects of love charms—ailments not recognized by biomedicine and ridiculed by it. They sold African muthi such as animal fats, which biomedicine claimed had no medicinal value, and bangalala, a remedy for sexual dysfunctions whose treat-ment chemists generally disparaged.[101] When asked about the selling of tokoloshe fat, Mr. Pimm claimed he thought the fat came from "some exist-ing animal in Portuguese East Africa."[102] He also professed ignorance about bangalala and its use as a male stimulant. The design of the Kwa Ndhlulamiti pamphlet—not only its Zulu title and subtitle ("The Chemist of Black Peo-ple") but also the accompanying picture—implied that the company was African owned. The picture was clearly meant to appeal to the variety of Africans coming into the cities. At its center stands an African man wearing a doctor's jacket and holding up a bottle of medicine. He tells a group of men, some in traditional Zulu headrings and others wearing t-shirts and suspenders, "By taking my medicine you will be cured quickly." The pamphlet appealed to traditional ideas of strong medicine, reading: "Only strong men can come with powerful medicines."[103] In essence, the pamphlets of white chemists looked very similar to those of their African counterparts; written in the ver-nacular, both sought to woo a new and emerging urban African population. By the early 1940s, an increase in mail-order patent medicines from "native chemist shops" could also be seen in places like rural northern Zululand.[104] Whether produced by white or African sellers of muthi, both attempted to legitimate themselves by appealing to African public perceptions of both bio-medical and African therapeutic ideals and authority.

Despite their increasingly precarious status, Natal's inyangas managed to maintain their legal standing, albeit a much weakened one, by exploiting the tensions and conflicting interests of the biomedical profession and the provin-cial government. Local officials, many of whom were firm believers in cus-tomary law and segregation, felt strongly about maintaining inyanga licenses. Many magistrates pointed out that biomedical doctors were inaccessible and too expensive for African clients, and that, either way, Africans preferred their own doctors. Others argued that repealing inyanga licenses would jeopardize relations between Africans and whites. Many municipal and magisterial au-thorities were confused by the type of license required for healers; some authorities issued traveling passes for the selling of African herbs, while oth-ers gave out general dealer's licenses. Furthermore, inyanga licensing enabled local governments to claim that Africans had access to healthcare in accor-dance with Natal's ideas of separate development.

PROFESSIONALIZING AFRICAN MEDICINE

To face the onslaught of legal challenges that began in the late 1920s, African inyangas like Bramwell Sikakane and Mafavuke Ngcobo joined together in

1931 to create the Natal Native Medical Association. This organization drew its support from the elite of the African inyanga community—mainly those who set up franchises, chemist shops, and mail-order businesses. While the association did not include the majority of healers (the unlicensed, those practicing in the rural areas, and Indian traders of muthi), most inyangas did benefit from the association's lobbying efforts to protect their occupation.

Hoping to win government recognition of their organization, the association sought to "professionalize" African medicine by using many of the same tactics as their white counterparts. Professionalism meant organizing themselves into an elite group that monopolized a distinct body of knowledge, enforcing codes of conduct, and, most important, convincing government officials that it was in the interest of the general public to protect their practice and status through legislation. Furthermore, the Natal Native Medical Association sought to convince the South African government of its unique status and to grant it a monopoly on "the native curative system" over healers not affiliated with the association.[105]

These goals were accomplished in a number of ways. First, professionalism demanded class and status. The Natal Native Medical Association hired lawyers, lobbied the national and provincial governments, wrote a constitution, held public meetings, and issued certificates to its members.[106] At a time when inyanga licenses proved hard to obtain, certificates not only conferred legitimacy to the holder, but also provided a useful means of raising revenue for the organization. Like their white competitors, the inyangas formed powerful alliances in an effort to influence government. Composed of an educated elite class of healers, the Natal Native Medical Association enjoyed the support of local chiefs, ministers, and teachers. Qandiyane Cele, the president of the Natal Bantu Minister's Association, wrote to the Chief Native Commission in support of this organization: "We can not say that they are better than white doctors, but we do say that they understand the many causes of black people's sickness than the white doctors and they examine their patient thoroughly and see if the patient needs treatment from the European side or native side." The Natal Native Medical Association also persuaded members of the Natal Native Congress (which later merged with the African National Congress) as well as John Dube (editor of *Ilanga lase Natal*) and A. W. G. Champion (Natal president of the ICU) to petition the government in its favor.[107] While this Christian elite had initially opted to assimilate to white South African culture and reject African "superstition," segregation and a decline in political and economic power encouraged a reassessment. The new African nationalism that emerged in the 1920s supported not only the Zulu king (a largely symbolic role after 1879) but African culture and medicine as well.[108]

Membership cards served to distinguish elite healers from those who could not obtain official licenses, particularly as they became more difficult to procure. Copy of "License for a Native Medical Practitioner." Pietermaritzburg Archive Repository, 1/HWK,3/1/10, HK 131/1902 (*top*); copy of a membership card for the Natal Native Medical Association. Pietermaritzburg Archive Repository, CNC, 50A (*bottom*).

Second, professionalism involved "formal" education. Knowledge of herbs, which had once been passed down through families, was transformed into a formal training program, one available to those who could afford it. Testifying at Sikakane's trial, Secretary General Ngcobo explained that all applicants served an apprenticeship of five years with a fully qualified and licensed herbalist. After such service, the Natal Native Medical Association's board of

SPECIMEN CERTIFICATE.

AFRICAN NATIVE DOCTORS' ASSOCIATION.

Tsoha
O Idirele

Vuka
Uzenzele

CERTIFICATE.

This is to Certify that..
is a Native Doctor and as such is a properly registered and approved member of the African Native Doctors' Association, by whose authority this certificate is granted.
This Certificate is valid for one year from date of issue.

Signed this.........................day of..........................194......

at... .. ....

....President.

Enrolled as a Member on the..

day of...194......

The Dingaka Association, renamed the African Native Doctors' Association, emerged in 1928. It sought to gain recognition for healers not recognized in Natal or the rest of South Africa. Membership cards not only provided the bulk of funding for the association, but given the lack of government licenses they served as a marker of legitimacy for the paying public. Specimen certificate of African Native Doctors' Association, as shown in the Constitution of the African Native Doctors' Association (Kroonstad, 1940). South African National Archives Repository, NTS, 7247, 204/326.

examiners tested a candidate for general knowledge and on the symptoms of and cures for specific ailments. Once a candidate demonstrated his full knowledge of the "native curative system," the board issued a certificate asserting the inyanga's exceptional talents.[109] In courts and in letters to the government on behalf of its members, the Natal Native Medical Association argued that its herbalists were better qualified than the average government-licensed herbalist who "picked up a smattering of herbs and does the rest by guess work."[110]

Third, professionalism needed a scapegoat. Using rhetoric similar to that of white medical practitioners, members of the Natal Native Medical Association defined themselves against the isangoma or "witch doctor" who "merely relies on superstitions of the Native race and on love-philters for his remedies."[111] Ngcobo described the inyanga practice as a "profession"[112] and emphasized that the organization aimed to preserve the "ancient art of Bantu healing from the degradation by quacks."[113] Mr. Ngcobo said the "object of the association was to preserve their rights and to cooperate with the government to bring to book impostors and quacks."[114] The Natal Native Medical Association's positioning of "genuine" versus "disingenuous" medical practitioners was an appropriation of the methods used by its white contemporaries to establish medical authority. As the majority of isangomas were women, the association in essence was acting to reserve the domain of state-recognized African healing for elite African men. But this move toward exclusivity was also in recognition of European lawmakers' obvious scientific bias, which had that led them to license herbalists and midwives over other healers in the 1890s. These strategies in Natal are unlike those of healers in the Transvaal, who organized all those who had been criminalized, including male and female healers of all types.[115]

Ngcobo's pamphlet expressed the need for culturally appropriate remedies for an African population and moreover the need for African healers to treat African patients. In his flier, he assured his readers that he was "a person who has acquaintance, knowledge and experience about various diseases of Bantu people, and is a Native (muntu) himself. At the same time the evil inventions or evil concoctions, charms and prevention of evil doings are known by a Native himself. I declare to all the Bantu people of South Africa that I possess medicine for all parts of the body and doctoring to all diseases that trouble a Native, starting from a baby, up to an aged woman and her aged husband. The medicines are specially prepared for Native troubles, custom and nature."[116] Beyond the appeal for culturally appropriate treatment lay a veiled use of racial rhetoric. African healers, like other members of the African elite of their time, used such imperialist rhetoric and also argued for segregation: white doctors for whites and African doctors for Africans, separate but equal, yet each a privileged group in society retaining status, wealth, and legal advantage.[117] As such, licensed healers argued that Indians and whites who secured general dealer's licenses should not be allowed to sell African medicines.

It can thus be seen that the men who established the Natal Native Medical Association used ideas of Western professionalism and appealed to African "tradition" as it suited them. They argued for government recognition as specialists and used the same arguments as whites to argue against the granting of general dealer's licenses to herbalists because they were "compounding liquid medicines from these herbs, practically in the same way as an apothecary."[118] On the other hand, the association did not criticize persons such as licensed inyanga Israel Alexander, who owned a mail-order business and storefront muthi shops that closely resembled those of white chemists.

Association members emphasized that while they might use the tools of biomedicine, their work was distinctly different from that of white doctors. As Secretary "Hlatswayo" (a.k.a. M. Ngcobo) stated, "By using these [stethoscopes, thermometers, and preservatives such as alum] we do not endeavor to mislead persons that we are qualified medical practitioners in the European sense. We advance with the times. As an illustration, we put our medicines in bottles of preservatives, whereas our forefathers put them in clay pots."[119] Elite inyangas had little interest in merging with biomedicine, and collaboration between the two groups would not have been an option at the time. As it is today, there are very real concerns over how collaboration can work with two very different ideas of wellness, as well as the government's continuing to privilege biomedicine over African and other therapeutics.

In essence, professionalizing African medicine meant clarifying what did and what did not fall within the boundaries of "tradition." As older men and chiefs used notions of "tradition" and the codification of customary law in the late nineteenth and early twentieth centuries to secure positions of power over women and young men,[120] elite healers worked to protect their healing methods through government recognition of their association in the hopes of gaining a commercial advantage over other healers and protecting themselves from aggressive medical legislation. In both cases, ideas of what constituted custom and tradition were stretched and imagined in such a way that they would benefit those who envisioned them. For these imaginings to come to fruition, however, required access to power. Historians Leroy Vail and Ranger show that the creation of customary law was the result of colonial interactions with "cultural brokers" or indigenous elites. In the case of elite healers, however, the biomedical lobby and their interest in checking the practice of healers prevented this type of collaboration with the government. Furthermore the government had little incentive to privilege an elite group of healers who provided no labor to the colonial economy and were successful and in some cases extremely wealthy businessmen. According to a *Sunday Times* article, I. Alexander's business was said to produce a daily sum of £200, and was sold in 1946 to Sampson Bhengu for £100,000.[121] The success of such healers clearly crossed a line that made whites uncomfortable and threatened the social, economic, and political order of white rule in South Africa.

CONCLUSIONS

Competition between inyangas and white physicians and chemists in Natal began when inyangas were first licensed but increased as inyangas began to enter the urban areas and develop their occupation commercially. This competition is evidenced in letters of complaint written by white medical practitioners to the Natal Medical Council, the Department of Public Health, and medical journals, as well as by their spearheading of legislation that limited African healers' practices and access to patients. As inyangas encountered other medical practitioners, African healers adapted their practices to include diagnosis, the use of new medicines and medical tools, and the advertising of their wares and services. Close contact between the African and white populations not only heightened competition between different types of medical practitioners but also prompted the biomedical community to seek legislative protection. Less willing to admit professional concerns over competition, the European medical community attempted to persuade the government by using rhetoric that highlighted concerns about racial contact and white "racial degeneracy." This rhetorical move was particularly important in gaining the government's favor and establishing biomedical authority in South Africa in the early twentieth century. Some African inyangas responded to biomedical competition, rhetoric, and the litany of legislation it created by professionalizing their own practices and seeking legal recognition for their organizations. Elite inyangas emphasized that their medical authority was derived not only through government-issued licenses, but through rigorous training and testing by their healing association. Though members of the Natal Native Medical Association never gained the official recognition of the South African government, they were instrumental in galvanizing support for their cause. As eloquent spokespersons and highly regarded members in their community, the association ensured that Mr. Bramwell Sikakane and other association members brought to trial for violating the Medical, Dental and Pharmacy Act of 1928 received only minor fines. Furthermore they provided inspiration and impetus to dozens of other healing associations throughout South Africa. Today that number has blossomed into more than 250 healing associations that, unlike their forebears, have been successful in their negotiation with the federal government over licensing and the legal standing of healers in the country.

African therapeutic encounters with biomedicine and medical legislation in the first half of the twentieth century effectively reversed certain transformations within African medical culture that had incorporated elements of biomedical practices. Culminating in the trial of Mafavuke Ngcobo in 1940, "traditional" African medicine found itself legally bound by white judges' definition of "native medicine." While "white" substances such as blue stone, castor oil, and Lennon's proprietary medicines continued to find a place in African therapeutics, those that too closely resembled biomedical ingredients, practices, or tools came under biomedical and legal scrutiny and were thus largely discarded. "Indian" substances and medicines, as we shall see in the next chapter, managed largely to avoid this same scrutiny and thus became incorporated into local African therapeutics.

5 ⇜ African-Indian Encounters and Their Influence on African Therapeutics, 1860–1948

In 1905, Parasoo Ramoodoo, an ex-indentured Indian living in Ladysmith, wrote to the Colonial Secretary of Pietermaritzburg to apply for a "native doctor's" license.[1] Ramoodoo's letter explained that he came from a long line of "native doctors" in India and had been practicing medicine for twenty-two years. He included a list of twenty-five patients, mostly Africans, whom he had cured, and asked that another "native doctor" subject him to an examination. Though the government had reputedly granted inyanga licenses to earlier applicants of Indian descent, Ramoodoo found his request—like others— denied on the grounds that he was not African.[2] Today's Indian practitioners of African medicine seem to face an equally liminal and dubious future, as predominantly African policy-makers assume that Indian healers and muthi shop owners are purveyors of goods rather than holders of indigenous knowledge. This chapter sets out to map one aspect of the largely unexplored and collective history of African-Indian encounters with regard to health and medical knowledge in Natal and Zululand.

Indians make up only a small minority of South Africans, approximately 3 percent, yet over a million Indians reside within the province of KwaZulu-Natal—home to six million Zulu-speakers. Within the confines of KwaZulu-Natal's cities and towns, Africans and Indians work, shop, play, and live within close proximity. Evidence of their encounters (historical and present) can be seen in news stories of Indian households terrorized by tokoloshes (the local African apparition said to be short, hairy and possessing a large penis) or of African schoolgirls suffering possession by *ufufunyane* (local African interpretations of Indian and/or white spirits).[3] Within the vicinity of Durban, one can

find Indian inyangas running muthi shops, Indians seeking out African healers and medicines, and African Muslims and non-Muslims visiting Muslim
healers and attending Sufi shrines for health and blessings. Indian muthi
shops as well as African sellers at the Warwick Avenue muthi market sell herbs
and medical substances of African and Indian extraction, both of which have
been incorporated within the body of African therapeutics. While rumors of
an African *poojari* (Hindu priest) at a Hindu temple and Indian isangomas
could not be confirmed, they are not outside the imaginations of modern day
Durbanites.[4] Indeed the cultural blending between Indian and African populations in a city that boasts of African rickshaw drivers bedecked in "Zulu"
beaded headdresses is hardly surprising. Such cosmopolitanism is not new
but has a long and complicated history with rural roots stretching far out into
the hinterlands of KwaZulu-Natal.

Despite an implicit understanding that many in the general public have of
the intermingling of African and Indian cultural ideas and practices, a certain
collective amnesia seems to have developed during the apartheid years. Present-
day antagonisms between Indians and Africans are seen by many as natural,
the result of two culturally distinct groups rather than a cultural production
arising from specific historic circumstances. Likewise, the general history of
Indian-African relations has received little attention from either scholars or
journalists. With the exception of a few dissertations and cursory paragraphs
in Indian–South African histories, little is written about their interaction.[5]
What is written mainly focuses on the 1949 Cato Manor riots, when Africans
destroyed Indian shops and homes.[6] Historical works on Indian South Africans
tend to be singularly focused, exploring the maintenance of Indian culture
and Indian-white relations in the struggle for social, economic, and political
equality. The post-apartheid era shows signs of a shift away from this earlier
historiography and toward an examination of Indian and African encounters,
albeit with slow beginnings.[7]

By examining the historical encounters of Indians and Africans around issues of health and healing in South Africa, this chapter seeks to demonstrate
the nature of medical pluralism found in Indian and African communities as
well as broader questions regarding their intercultural contact. The purpose is
not to quantify such encounters, but to show a pattern of mixing. This chapter provides evidence of polyculturalism in both groups between the 1860s
and late 1940s, but particularly the polycultural development of African therapeutics. By delineating a few threads of influence between Indian and African
therapeutics, this chapter explores why certain cultural ideas and practices with
regard to medical therapeutics were adopted and incorporated while others
were rejected. It also asks how these medical knowledge systems, practices,

and artifacts were transferred, produced, and negotiated between these communities with different interests, cultural beliefs, and degrees of access to power. Last, it shows how cultural actors, who asserted and assigned identity for political, social, or economic gain and patrolled cultural—and in this case medical—boundaries for transgressions helped to influence the polyculturalism of African therapeutics.

Before I begin let me touch on some basic problems that arise from conducting research on the historical encounters of these two communities in South Africa. The main issue, as we have seen in some of the other chapters, is availability of sources. Missionaries, government authorities, and anthropologists writing during the late nineteenth and early to mid-twentieth centuries generally wrote about specific African or Indian communities. None seem to have had much interest in African-Indian relations, and this is reflected in their writings. This lack of interest was combined with reluctance on the part of healers and patients to reveal information on the marginally legal, potentially taboo subject of health and healing. Consequently healers left relatively few archival records behind. Interactions between Indian and African healers and their patients are thus difficult to trace, and it seems reasonable to assume that for every Indian inyanga that appears in the archives, there are more like him who purposely escaped the attention of government officials. Likewise, in the early 1950s, anthropologist Hilda Kuper observed that Indians she interviewed in the Durban area had a desire to appear "modern" before their white interviewer and thus were often reluctant to admit use of Indian healers, let alone African ones.[8] Consequently, archival records of Indian-African encounters usually show up in the form of complaints filed by individuals, or incidentally. This lack of evidence regarding group interaction, however, should not always be taken as evidence of absence.

In this chapter, I have pieced together bits of evidence from numerous sources from a hundred-year period. Interviews with a number Africans and Indians, most of whom were (and are) involved in the practice of healing, help to fill out some of the historical details, but also tempt one to project present-day phenomena and concerns onto the past. Unfortunately it is difficult to gauge an exact time-depth to some of the phenomena described. Consequently some conclusions in this chapter remain speculative and await further research for confirmation.

WIDER AFRICAN-INDIAN ENCOUNTERS

Indians and Africans have been interacting in Natal since Indians first arrived as indentured servants to work the sugar fields that line its coast. It was here that Indians and Africans first encountered each other. Between 1860 and 1911, 152,184 Indians (from current-day India and Pakistan) entered Natal as indentured laborers. In the 1870s Indians composed 42 percent of the labor on sugar estates in Natal, reaching a maximum percentage of 87 percent in 1907–8, but dipping down to 17 percent by the 1930s.[9] The majority of non-indentured sugar workers were recruited from neighboring Mozambique and Pondoland, as estate owners favored Africans from outside the borders of Natal for long-term labor contracts. Africans in Natal preferred to work their own land and generally required higher pay than outsiders. By the 1930s, however, rural disintegration, poverty, and South Africa's land policies had forcefully pried local Africans from their land. Estate owners obtained some of this freed African labor by successfully lobbying the government to limit mine recruitment from Zululand.[10] A lack of statistics on the local labor supply, however, makes it difficult to determine the exact number of Zulu-speakers who may have been hired on a temporary basis.[11]

African and Indian encounters that originally took place in the sugar fields[12] continued as the system of indenture spread to other industries and free Indians moved into the rural areas to take up agriculture. The indentured system itself not only led to some social intermixture, but most likely led to a degree of miscegenation between Indian men and African women,[13] as Natal recruited a hundred Indian men to every thirty Indian women, later increasing this ratio to two to one.[14] At the turn of the century, Stuart's respondent Qalizwe tells him that he has heard of such sexual relations in areas where Africans and Indians worked and lived together as equals.[15] Other Indians joined African communities to escape the degradation of indenture,[16] while some free Indians chose to live under African chiefs[17] and at least one Indian leprosy sufferer chose to live at an African leprosarium rather than return to India.[18] Africans and Indians (both free and indentured) worked together in the construction of railroads (1876–1911), in the coal industry (1876–1911), and on small family farms. They also interacted in urban areas like Durban and Pietermaritzburg, where they resided in ethnically mixed communities until the mid-twentieth century.[19]

A second wave of Indian immigrants known as "passenger" Indians—they had paid their own way—came to South Africa starting in the 1870s to service the Indian indentured and ex-indentured communities. Many Gujaratis as well as Muslims from the broader Indian Ocean world arrived during the late nineteenth and twentieth centuries. "Passenger" Indians made up the bulk of the middle to upper classes, and many became successful merchants to both the Indian and the African communities.[20] This population group, however, made up only about 10 percent of all Indian immigrants.[21]

In 1881, twenty years into the indentured system and ten years after the arrival of the first "passenger" Indians, the acting secretary for native affairs described the nature of African-Indian relations as follows: "They [Africans] have become reconciled to them [Indians] in late years, but at first they despised them. They seem to fraternize now. They [Africans] looked upon the introduction of the Coolie [Indian] with some little apprehension at first, but they now understand it was owing to the scarcity of labor."[22] In that same year, an African living near Verulam echoed a similar sentiment, stating, "We get on well with Coolies, they stay at our kraals."[23] Within the mixed housing units of Bamboo Square that lined Point Road in Durban, Africans were said to be quite fond of Indian *dahl* or *dali*.[24] As further evidence of Indian-Zulu interaction one need only observe the importance of Fanagalo, a linguistic mixture of Zulu, English, and Afrikaans that served as the workplace *lingua franca* among Indians, Africans, and their white employers until the 1930s and '40s. Among early Indian immigrants, more were familiar with Fanagalo than English, and sometimes Fanagalo served as a common medium between Indians of different language groups.[25] Not all intercultural relations between Indians and Africans were positive, particularly where there were stark class differences such as in areas like Umzinto in the late 1890s, where Indian farmers bought up land and forced Africans off their own. In such instances relations soured and each group blamed the other for local problems.[26]

By the 1890s some Indians and Africans began to make political alliances, even if loosely conceived. For instance, much can be made of the friendship between John Langalibalele Dube and Mohandas Gandhi, whose Indian National Congress (1894) influenced the formation of Dube's own Natal Native Congress (1900), precursor to the African National Congress (ANC). Gandhi wrote about Dube in his publication *Indian Opinion* and shared his printing press so that Dube could print his own newspaper, *Ilanga lase Natal*.[27] Initially the political elite of both communities decided they had different struggles and needed to work separately.[28] By the late 1930s, however, Indian and African elites recognized a common fate as "nonwhites," and at the prodding of the working classes these communities began to organize together. A series of nonracial, nonethnic political groups began with the Non-European United Front in 1939, followed by the Non-European Unity Movement in 1943; the Doctors' Pact between A. B. Xuma, G. M. Naiker, and Y. Dadoo in 1947; the Defiance Campaign in 1952; and the Congress Alignment in the 1950s. In 1955–56 the Indian newspaper *The Opinion* regularly published "The African Viewpoint," penned by a Mr. Ngubane.[29] Such cooperation aimed toward a common multiracial, multiethnic, democratic struggle at both the local and the national level. In 1946 the Coloured community and the ANC sent delegations to Durban to protest the Ghetto Act, which exclusively targeted Indians. And in 1947 the South African Indian Congress helped send Dr. Xuma to the United Nations in a campaign against South Africa's incorporation of South West Africa.[30]

An important part of this cooperation came from below. Prashad argues, "Polyculturalism exists most vividly among the poor and working class, among people who are forced to live among one another and who ultimately work together towards freedom."[31] This particularly seemed to be the case in other British colonies—Jamaica, Trinidad, and Guyana—where indentured and ex-indentured Indians, rather than the merchant class, made political alliances and mixed socially with other population groups.[32] Where Indian immigrants came largely from the merchant classes, such as in Uganda, Kenya, and Malaya, cultural intermingling with the local population occurred less frequently.[33] All diasporic Hindus during the early period experienced *kala pani* (dark waters), the long ocean voyage in which proper caste obligations were violated and religious purity lost.[34] Yet wealthier Hindus, particularly those of higher castes, had more incentive to recreate caste divisions once within the diaspora than did lower- or unscheduled-caste Indians. Within South Africa the majority of Indians arrived as indentured laborers, with 60 percent from the lowest and unscheduled castes. Living and working conditions of indenture within Natal made reestablishing or maintaining caste obligations impossible and thus lessened the importance of caste in these communities.[35] Early intermingling between working-class Indians and working-class Africans may have been a means for Indian survival or social mobility, but the decline of caste certainly eased those relations in a way unlikely to have occurred with the passenger class.[36] Connections between working-class Indians and Africans led to a quick and bloody resolution of the 1913 Natal Indian Strike when rumors circulated that Africans planned to join in with their working-class Tamil allies.[37] This class solidarity between working-class Indians and Africans was prompted in part by their alienation from the upper classes, which pursued financial benefits and political recognition for themselves.[38]

A number of scholars working on Indian or Zulu communities have assumed that cultural differences between these groups—a long literate tradition versus an oral one, largely monogamous and extended families versus polygamous ones—would bar their interactions.[39] While there were/are certain differences, there were also a number of similar cultural ideas and practices, particularly regarding issues of health and healing, that may have eased encounters between these two communities. Some of these similarities, particularly with Hinduism and Zulu religious practices, included similar rituals performed at birth, a division between herbalists and healers who used spirit possession, a belief in witchcraft, spirit possession of the afflicted, faith healers who used holy ash, the use of enemas and emetics, and a shared skepticism of biomedical doctors and hospitals born out of their forced subjugation to a white and biomedically dominant society. The exchange in medical cultures described below developed largely out of interaction between these two working-class communities.

INDIAN THERAPEUTICS IN NATAL

Before examining the encounters of Africans and Indians with regard to African therapeutics, it is important to first visit the practice and use of "Indian" therapeutics in Natal. This will enable us to gauge the availability and possible transfer of medical techniques and substances into African therapeutics. Given that South Africa's Indian population was quite heterogeneous, it is not surprising that they also practiced a wide variety of different types of therapeutics. While it is difficult to gauge how many of these practices were in existence during the nineteenth century, we know that in the twentieth century many persons classified as "Indian" still retained knowledge of household remedies from the subcontinent and continued to consult Hindu, Islamic, and Ayurvedic spiritual and medical practitioners. This is despite the fact that indentured Indians, in particular, were subject to the scrutiny and imposition of biomedicine more than any other South African population group. The use of Indian cures and practitioners persisted through to the 1950s and in some communities continues to the present day, with the majority of healers being Indian spiritual healers.[40] Indian approaches to health were numerous and passed down through family members and re-enforced through the ongoing influx of new Indian immigrants that continued throughout the nineteenth and twentieth century, though indentureship ended in 1910. The use of household remedies and the consulting of Hindu and Islamic spiritual healers proved the most accessible and popular means of remedying illness. Few *vaidya* (practitioners of Ayurveda), however, seem to have survived from the period of indenture, and current vaidya have usually trained in or have recently emigrated from India. Consequently "professional" Ayurveda has begun reasserting itself only more recently in South Africa.[41]

In the early to mid-1950s, anthropologist Hilda Kuper and her research assistants interviewed hundreds of Indian women and men about their day-to-day lives in the Indian communities of the larger Durban area. Much of the information collected revolved around issues of health and religion, which consequently became the main foci of Kuper's book *Indian People in Natal* (1960). These interviews showed that Indian South Africans in the 1950s retained many remedies and rituals from India, demonstrating many similarities to as well as some continuing differences between practices of Tamil/Telegu and Hindi speakers. Kuper's book details various interpretations of illness by these communities and the strategies used to remedy such illness. Her research also showed that Indians maintained access to medicinal herbs from the subcontinent by growing them in their gardens and buying them from Indian shopkeepers. Indian herbal remedies such as honey, castor oil, cloves, ginger, garlic, mother's milk, betel leaf, areca nut, turmeric, and syringa leaves were used in numerous therapies as well as health rituals surrounding birth, menstruation, and marriage.[42] A popular Indian cookbook created by the Women's Cultural Group for the "South African Indian housewife" in 1961 lists twenty herbs grown from days of indenture under the title "medicinal spices," and a number of others under a section entitled "convalescence and remedial foods."[43] Likewise interviews with Indian South Africans in 2002 showed a continuation of many of these same practices and of the use of medicinal herbs. Kuper's surveys in the early 1950s concluded that Indians spent more on patent medicines than on hospital and doctors' fees; it is unclear, however, if these were Ayurvedic medicines packaged in India or Western patent medicines.[44]

In Indian as well as local African cultures it was and is not possible to separate health from religion and social well-being. Indian South African communities attributed illness to a number of sources. The nature of the illness and circumstances surrounding its emergence determined how and from whom treatment was sought. Illness could be attributed to natural causes, retribution for sins in a former life, a visitation from a deity, losing faith in god, neglecting a Hindu house spirit, witchcraft, possession, or the evil eye. Consequently there were numerous strategies for regaining health. While patients might be Hindu, Muslim, or Christian, it seems that there was some crossing over of both ailments and treatment—the diagnosis influencing the practitioner sought. Kuper's research assistants note that a Christian afflicted by a Hindu deity would attend a Hindu temple or seek a Hindu practitioner for relief. Often, however, a number of different types of practitioners would be used simultaneously. In many instances a patient was first brought to a temple or mosque for diagnosis and treatment. By the 1940s and '50s, however, biomedical health education regarding the germ theory meant more people also began to seek the inexpensive treatment of local government health centers. Many Indians, however, remained skeptical of biomedicine even in this later period.

While there are a number of different types of Hindu practitioners, Kuper's book focused predominantly on the spiritual healing performed by a poojari

(Hindi) or *poosali* (Tamil) of a Hindu temple who healed through his or her possession by a Hindu god or goddess with little recourse to herbs.[45] Other strategies for ill health included promises to "carry *kavady*" or trance possession of a deity during religious holidays in exchange for one's own wellness or that of a family member.[46] The Muslim community also had recourse to spiritual healing by Islamic healers who could help with everyday illnesses or illnesses resulting from influence of *jinns* or the evil eye. Depending on the type of Islamic healer, one might be offered herbal remedies, prayer, exorcism, or astrological readings. Furthermore, many Indians, regardless of faith, often visited local Sufi shrines for blessings and wellness.[47] In the case of temple or mosque healings, practitioners relied mainly on the use of holy ash and water to cure sufferers.

The historical attrition of vaidyas and *hakims* (doctors of Islamic medicine) and more recently the declining use of Indian household remedies resulted from: biomedical coercion practiced by the indenture system; the South African government's refusal to acknowledge Ayurvedic practitioners in Natal during the late nineteenth and twentieth centuries; and more recent influences of Christianity, Western-style education, and the changing importance of and ideas concerning "modernity." Interestingly some of today's Indian inyangas descend from an Ayurvedic practitioner who came to South Africa as an immigrant in the nineteenth century.[48]

AFRICAN-INDIAN MEDICAL ENCOUNTERS AND THE RISE OF INDIAN INYANGAS

In a medically plural society like South Africa, both whites and Indians explored the efficacy of African therapeutics, and in some cases, like Ramoodoo's, became practitioners themselves. A few white chemists sold African muthi, though it was generally obtained through the post, which ensured a degree of social and physical distance between them and their largely African clientele. Indians, however, became not only successful inyangas, but itinerant merchants of African medicinal herbs and eventually owners of muthi shops. Indian healers not only learned Zulu, but also lived and worked among the African population. Far from being socially stigmatized, Indian inyangas, who became more popular in the mid-twentieth century, enjoyed high standing. A 1950s survey of the Indian community showed that Indian inyangas ranked in the highest status group alongside principals, teachers, and medical aides.[49]

Indian healers first emerge in the archival records in 1881 in the Durban vicinity. In that same year, a medical officer claimed African cases of syphilis were being "treated by the various druggists and the native and Indian so-called doctors."[50] By 1892 and 1893, Indian healers were spotted in the outlying In-singa and Umgeni districts, "traveling about the locations" and doctoring Africans.[51] By 1898, an "unlicensed" Indian healer who poisoned an African chief he treated for an eye disease came to the notice of the Natal Medical Council.[52] By 1905 Ramoodoo's plea from Colenso further indicated that the phenomenon of Indian healers now existed province wide. Though sporadic, these early archival references indicate that Indian healers provided some sort of medical therapy to the African population. What is less clear is whether In-dian healers used Indian or African therapeutics or a combination thereof.

While we might assume that Ramoodoo, by virtue of his inyanga applica-tion, had been using African medicines since 1883, specific evidence of In-dian healers utilizing African therapeutics emerges in 1908 in the rural area of Krantzkop. An African police informant reported that an Indian healer named Masibawezinja "went to Mapukwana Gwananda . . . and charged him ten shillings for cutting him and putting in medicine *(gcaba).*"[53] The informant's parenthetical *"gcaba"* points to a local African form of administering medi-cines, but this healer's Zulu name, which literally translates as "dogs' feces," indicates some tension and his inability to win over all of the local African population. Oral testimony from present-day healers also points to the emer-gence of Indian inyangas at the turn of the twentieth century. L. Govender (who used to own an Indian muthi shop in Durban) said his grandfather gained his knowledge of African therapeutics from Africans he worked with in the coal mines, but also from an uncle already in the "herb business."[54] Mata Venkata-chellam Naidoo, another Indian inyanga who began his occupation at the turn of the century, learned about African herbs while bartering animal skins between blacks and whites as he traveled throughout Natal, Lesotho, and Swaziland.[55] "Masibawezinja's" example, coupled with oral histories, indicates that some Indian healers had begun using African therapies and that others considered themselves inyangas at the beginning of the twentieth century.

While it is unclear why Indians became inyangas, at least some them, like Mata Naidoo and Ramoodoo, had been vaidyas[56] or came from families of this tradition. This hereditary aspect of healing is common among both African and Indian healers. The descendants of Mata Naidoo explained that their grandfather carried on the Ayurvedic tradition as he found African herbs re-sembling those found in India. T. Pillay elaborates:

> It was my mother's father that brought herbs from India, Ayurvedic herbs and mixes it with stuff from here. And we then short of it, and we go out and looked for it in the veldt [field]. And looked for it, be-cause some of the roots were found in India—and I can show you

some roots here that were found in India and found in Africa. So the continent could have been at one time, together, and being an herbalist or Ayurvedic practitioner he could find these things in the veldt, and the leaves and things and take this herbs and put together and get it; so from him my mother learned and my uncles learned.[57]

A botanist at the Durban botanical garden, as well as an Ayurvedic practitioner trained in India, confirmed that a number of genera are common to both India and southeastern Africa.[58] Thus Indian vaidyas in Natal acted much like early European chemists at the Cape who sought to replace more expensive remedies from overseas with less expensive and locally available African herbs.[59]

Indians not only practiced African medical techniques and used African herbs in the late nineteenth and early twentieth centuries, but also began to actively engage in the medicinal trade. The ecological crises of the 1890s and early 1900s, combined with the social crises accompanying increased taxation and migrant labor, led to rampant physical and social illness. As African healers began to travel the countryside offering therapeutics and herbal remedies, so too did Indian healers, though there is no archival evidence for this until the early twentieth century. "Masibawezinja," the unpopularly named Indian healer, was one such itinerant. Reportedly carrying a large quantity of medicine, he was accompanied by an African and sold medicines from kraal to kraal. Apparently he refused African customary forms of medical payment such as a livestock and insisted on cash.[60] Likewise in 1914, two Indians who canvassed African patients in the areas of Thornville and Verulam were found guilty of practicing medicine without a license and sentenced to a fine of £20 and three months' imprisonment.[61] Mata Naidoo also seems to have been a traveling healer until the 1950s, though one who accepted more conventional forms of payment. As another grandson explains, "My grandpa he worked from home [Newcastle], was what you call an *inyanga*. He would go away from home for months at a time traveling only by horseback and come back with lots of things, herbs, sheep, cattle — that is until people began to pay in cash."[62] These long journeys, coupled with numerous payments, point to the practice of canvassing or touting remedies rather than traveling to attend one's patients, as was the practice of earlier African inyangas. Less ambiguous is L. Govender's description of his father and grandfather's occupation as "a chemist on wheels." The Govender men would leave for months at a time, hawking herbs by horseback and wagon in the Drakensburg Mountains and other rural areas of Natal. They did not buy herbs but dug them up themselves. Govender says, "They usually went to Basothu people, to cure the people. They were like traditional healers—*izinyanga*." But then adds, "No, they didn't wear traditional *inyanga* attire. They didn't throw bones." Rather, they made medical recommendations based on people's complaints and responses to their questions. Other Indian inyangas, however, did "throw the bones" and were sought after by such high-ranking Zulus as King Solomon to determine a case of witchcraft.[63] This medical trade, as Govender implies, was an important precursor to the "chemist" or muthi shops of the 1930s and '40s.[64]

Seeking to avoid prosecution and to ease transit around the province, some Indian healers sought legal recognition with an inyanga's license. Government officials often stopped and demanded paperwork from Africans and Indians as a means of ensuring compliance with labor laws. Those who could afford it thus sought out licenses. While Ramoodoo appears as the first Indian on record to apply for an inyanga license, he believed that other Indians had obtained licenses before him. Initially the application for an inyanga license required the support of a local magistrate and African chief, presumably to deter itinerant healers and ensure a level of accountability.[65] In 1913 "Indian Seetal," who lived near Wasbank under chief Ncosana, persuaded this same chief to support his inyanga license before the Dundee assistant magistrate. Seetal explained that he had lived among Africans for eighteen years, learning to become a "native doctor and herbalist."[66] Likewise two other Indians, Chinnavadu in 1917 and Sevuthean in 1918, applied for inyanga licenses with the support of African chiefs.[67]

Like Ramoodoo, these Indian inyangas found their applications denied by the chief native commissioner on the grounds that licenses were issued only to "natives." Furthermore, they were informed that no provisions existed for Indian medical practitioners. White administrators remained almost uniformly unmoved by the background of the applicant, regardless of how long he had been living with the African population or his degree of cultural assimilation. Rather, decisions were made based entirely on the administrator's perception of the applicant's race. Thus in the twentieth century, Indians, as well as a few whites and "coloureds," were largely denied inyanga licenses, while Africans denied licenses outside of Natal found their applications accepted within the province. Indian inyangas learned to get around this legal restriction by applying for hawking, and, later, general dealer's licenses.

These application cases are particularly interesting on several grounds. First, they demonstrate a process of cultural exchange in which some Indian healers learned African therapeutics and used them with African patients. Second, they show that Africans and African chiefs were willing to support some Indian healers during a time when many Africans were loath to see biomedical practitioners. And third, we see governments (that relied largely on tactics of division) reluctant to encourage cultural exchange. Certainly racial and cultural "transgressions" between the white and nonwhite communities proved alarming and would provide a basis for medical legislation of African medicine, but the sharing of cultural ideas and practices between Indians and Africans also signified uncomfortable possibilities of political unity against white rule. Vigilance over these nonwhite boundaries, however, was sporadic and did not deter Indian inyangas and the rise of fixed commercial premises from which Indians could sell African herbs and medicines.

THE RISE OF INDIAN MUTHI SHOPS

The Indian muthi shops that dot the urban landscapes of Zululand and Natal as well as the streets of Johannesburg all have a very similar look to them. Lined neatly on shelves or in cubbyholes one finds roots, bark, leaves, or popular dried flowers like *impepo* (used regularly by Africans to call upon the ancestors). In addition there are large bottles filled with the most exotic and expensive herbs, and many containing brightly colored Indian substances. Some shops offer patent medicines, including local concoctions like *Zifu Zonke* (literally, "all the diseases"), Ayurvedic medicines like senna leaf, and "Dutch" remedies like Lennon, which bottled European and African home remedies. Medicinal mixtures are usually made fresh on demand, wrapped in newspaper, and given to the customer along with verbal instructions if necessary. Sometimes store-made infusions contained in recycled bottles are sold, though technically this was illegal until recently. African workers do an odd assortment of jobs, chopping up roots and barks, grinding herbs into a fine powder, running errands, and occasionally serving customers. While these workers may be knowledgeable about African herbs, they are not inyangas, and all serious medical inquires that I observed were directed to the Indian inyangas. These shops operate much like biomedical chemist shops in South Africa, selling remedies but also occasionally offering unofficial diagnoses. In more recent years, some muthi shop owners have added various African artifacts—Zulu shields, *iwisas* (fighting sticks), ostrich eggs, and pictures or mountings of African animals. As one healer told me, he hoped such decorations might attract Western tourists in search of an "authentic" African experience.

As discussed in chapter 4, the rise of muthi shops accompanied South African urbanization, migrant labor, and the newly developed muthi markets.[68] Photographic evidence and descriptions from the 1910s show that Africans initially dominated this trade. Mr. Chittenden described the Durban muthi market in 1915 as follows:

An Indian inyanga, his assistants, and his shop in Pietermaritzburg, KwaZulu-Natal, 1998. Note the large bottles of powders (umkhandos) in the forefront. Photo by the author.

Indian muthi shop in Ladysmith, KwaZulu-Natal, 2002. Photo by the author.

I came out into a yard which proved to be the consulting ground of a number of native (Zulu) medicine-men licensed to practise. They sat in rows on either side of the yard with their multifarious remedies spread out on sacking before them. And what remedies they were too! Anything, varying from a small frog bottled up in spirits of wine, to the dried entrails of a crocodile . . . There were cochineal and tomato-sauce bottles filled with muddy-colored liquids and potions . . . there were pieces of bark, bones, gypsum, asbestos, felspar, crocidolite, corks, dried roots, calabashes, seeds, chunks of elephant hide, animals' claws, assorted varieties of dried skin, fragments of china.[69]

In Durban, the Warwick Avenue and Grey Street area described above became one of the busiest sites not only for informal trading, but also of Indian-African encounters.[70] African muthi sellers and inyangas situated outside the Municipal Eating-House found themselves among Indian hawkers and peddlers of fruits, vegetables, and small wares, each serving a largely Indian and African clientele.

Sometime between 1915 and the late 1920s the muthi trade changed to include Indian as well as African inyangas and muthi street traders, and in the 1930s a number of muthi shops opened. Braby's Durban Directories indicate that the first "native chemist" shops opened in 1930. One was opened by an Indian, Mr. V. Mahomed, on 129 Queen Street, while the other, on 37 Queen Street, belonged to an African, Mr. C. L. Dube.[71] In each of the following years, Indians and Africans opened and closed new herbalist shops in the Durban Casbah, listing themselves in Braby's as a "native chemist" or "native herbalist." In 1933, Dr. Cawston mentions that he had found "medicine men," both "native and Indian herbalists," selling remedies on the roadside and in the "native market," as well as a number of shops that sold "native remedies." "One shop I visited in a leading Durban thorough fare contained 250 bottles of medicine on the shelves; infusions and decoctions were being made . . . Hanging round the shop were numerous fresh bulbs, roots, branches, leaves and portions of bark, besides the skin of a porcupine and lizard and the skull of a small mammal, a dead bird and a sandshark."[72] Cawston's descriptions of herbal and animal remedies as well as premixed infusions are reminiscent of Chittenden's vignette some eighteen years earlier. What is different is the housing of remedies indoors rather than outdoors, as well the number of bottles, which probably contained ground herbs. In 1938 and through the 1940s, Africans dominated herbal shops in downtown Durban, spreading to Grey, Victoria, and Umgeni streets.

Muthi shops offered what many of the traders on the streets did not. They enabled sellers to better preserve their wares, but also to stock larger quantities, offer more varieties of herbs, and consult with patients in a more private setting. Many of the muthi shops had separate consultation rooms.[73] Furthermore a muthi shop appealed to local African ideas of *hlonipa* (honor and respect) for the way in which muthi was housed, a questionable aspect of street trading.[74] Part of the move toward shops rather than informal street trading in the '30s and '40s, however, may also have resulted from a law that prohibited selling herbs outdoors.

L. Govender's father, like many of the African informal traders, initially sold his herbs on the pavement in Victoria Street. After five or six years, when hawkers could no longer sell herbs on the street, he opened a shop, "Kwesinyama," a few blocks away at 156b Queen Street in 1937. His brother-in-law C. V. Pillay had opened a shop the year before at 131c Victoria Street and helped Govender in setting up his shop.[75] An African healer, P. Cele, opened a similar shop on Leopold Street in 1938 for the same reason.[76] The growth of muthi shops can also be seen in the number of general dealer licenses distributed to shop owners. By 1936, twenty-seven general dealer's licenses for selling herbs and "medicinal compounds" had been granted to Indians and Africans in the Durban and Pietermaritzburg areas.[77] Ten herbal shops are listed in the Casbah district itself, though at least three of these have European names. By 1945 the Casbah alone housed twenty-seven muthi shops or consultation rooms with proprietors of various ethnic backgrounds.[78]

Indians and Africans obviously influenced each other in the establishment of Natal's muthi shops. They both certainly borrowed the practices of the street hawkers and inyangas who came before them. "Kwesinyama," the black bull, named after the Indian owner who earned that title on the soccer field with African teammates, is one of the oldest muthi shops in Durban. Cele, an African shopkeeper, is said to have used the Kwesinyama shop as a model when setting up his own. Cele's son, a well-known African inyanga and muthi-shop owner in Umlazi, said, "Even now we display like the Indians. My shop is same, like an Indian shop." As contemporaries for thirty or more years these two shop owners "had good relations. They would buy from each other that which they didn't have."[79] Likewise a current Indian shop owner, speaking of her grandfather's era claims, "I don't think there was any competition. Basically, everything was just fine . . . Everyone knew one another and there was no back biting."[80] When asked the difference between African and Indian muthi shops, Rev. Mpongose told me in 1998, "They are the same, there is no difference. Because in those Indian muthi shops the medicines that are sold there are coming from all over, and then those Indians buy and then they have employed people who know about those medicines so they can serve the right stuff . . . People have faith in it because they are using our medicines.

And then in our medicines we are not using only medicine found in Africa; some are coming from India."[81] Current muthi shops throughout KwaZulu-Natal are said to resemble Durban's first Indian and African muthi shops. Gone, however, are the endangered flora and fauna that gained protection from the Natal Parks Board in the 1970s.

Indian muthi shop owners, however, had a distinct commercial advantage over their African counterparts. Though the white government restricted African and Indian purchase of property, Indians could generally obtain shops closer to the city or town center and, more importantly, near major transportation hubs. Proximity to such areas was often a predictor of commercial success. Yet as L. Govender states, "The blacks were not allowed to open up shops in urban areas. In European areas we were not allowed to take up business, but we used a white nominee. In the Grey Street area [which abuts Durban's train and bus terminals] it was okay."[82] This put Africans in a potentially difficult position. Under apartheid, Indians were allowed to own shops adjacent to the railway stations in both downtown Durban and Pietermaritzburg. Apartheid administrators forced African merchants out of the cities and into African townships, where they serviced a local rather than a metropolitan clientele. Cele and then his son purportedly owned the lone African muthi shop in Durban after the enactment of the Group Areas Act and managed to stay by renting from an Indian man. After years of struggling each December for an extension, however, the younger Cele left the city center in the late 1970s and eventually succeeded in Umlazi township.[83] The restrictions placed on hawkers and African merchants successfully minimized competition for Indian muthi shops. By the mid-1980s Indian muthi shops prospered to such an extent that the Sunday Tribune reported some one hundred shops in Durban's Grey Street area. Likewise with the end of apartheid and the revitalization of street trading and the Warwick Avenue muthi market, commercial competition has resulted in a decline of Indian shops, which incur higher overhead costs. By 1997 only fifteen Indian muthi shops remained in this area.[84] Referring to Durban's muthi market, which now operates in direct competition to the nearby muthi shops, Naiken laments that her customers "can get everything up there, not like before where we could sell the powders, and get the blue stone and sulfur and things like that we could sell, now they have everything."[85]

INDIAN INFLUENCES ON THE LOCAL PHARMACOPOEIA

In 1998, in Durban, I interviewed an African isangoma concerning the important role that patients played in shaping her own notions of "traditional" African therapeutics. She emphasized that in a multitherapeutic community, patients came to her for "traditional" healing, and thus she obliged by, for example, giving patients "traditional" herbal remedies in a chopped form rather than the pills now widely available via the growing "natural remedy" market. When I asked her to prepare me an intelezi (protective medicine), I noted that she included a sprinkling of *umkhandos*, or bright pink and yellow Indian powders, in the otherwise chopped herbal mixture. While this inclusion of umkhandos in an intelezi may have been unique to her, many African healers have incorporated Indian herbs and substances into their remedies. The inclusion of such substances reflects not only an Indian influence but also an African appropriation of Indian substances within an African etiological framework.

It is difficult to measure when exactly these substances came into the local pharmacopoeia, as the majority of evidence comes from present-day oral histories. Though a number of Indian and African inyangas mentioned that their older relatives had sold or used Indian remedies, I found only one archival reference—from 1940. Mafavuke Ngcobo's assistant testified in court to his using *hla kwa zaseIndia* (croton seed) with the local herb vutana (not listed) to create a mixture to remedy back, bladder, and stomach pains as well as dysentery. Senna also appeared as a court exhibit, along with other substances found in Ngcobo's shop.[86] The current acceptance of Indian herbs and religious powders into African therapeutics resulted not only from the day-to-day encounters of Indians and Africans, particularly before the onset of apartheid, but also from the influence of Indian inyangas and muthi merchants themselves. It is probable, however, that the rise and prominence of Indian muthi shops conferred by the Group Areas Act helped to speed up this incorporation.

As Indians explored the efficacy of African therapeutics, African healers and patients investigated Indian and biomedical medical cultures for their usefulness, sometimes incorporating these elements into African medicine and practices. One Indian muthi shop proprietor noted that Africans tended to use Indian remedies like senna and syringa because African mixtures "always work with their stomachs. They drink and they bring up, they *silinga* [derived from Zulu for syringa tree]."[87] A number of African healers whom I interviewed in 1998, both in Durban and in the rural areas of Mthubathuba and Hlabisa, noted other explicit reasons for using Indian substances (specifically patent medicines). Several healers mentioned that they used Indian herbs because they could read the instructions on the patent medicine box. Another healer mentioned learning to incorporate Indian medicines within his own mixtures from his parents and as a result of working in Johannesburg. Indian muthi sellers themselves were said to have instructed healers in the use of Indian herbs. Other African healers mentioned that Indian remedies proved

An Indian muthi shop in Newcastle, KwaZulu-Natal, 2002. Photo by the author.

particularly useful for a number of illnesses (both "culture bound" and non–culturally specific ones); these included *umpundulu* (an African illness caused by witchcraft), *idliso* (an local ailment caused by poisoning), and lice. *Isibeba* (Indian aloe), an Indian medicine mixed with Zulu medicines, helped to take out "pimples inside and outside the body," while *ujamalikwata* (another Zulu term for croton seed) worked as a blood tonic and laxative. While many of these remedies were available in pill form, most healers preferred to grind them, and many added them to porridge or muthi mixtures.

More widely used and less likely identified as "Indian" medicines *per se* were substances for Hindu religious purposes, Indian home remedies, and herbs and spices. Umkhandos originate from the brightly colored powders that Indian women used as religious ornamentation and that were used to decorate Hindu religious statues. Umkhandos (generic Zulu term for all of these powders, each color identified by its own Zulu name), most likely appealed to African healers due to their bright colors but also for their fine powdered consistency. In the early twentieth century, ground muthi held certain significance, being equated with powerful individuals who used it to gain power over others. An inyanga interviewed by James Stuart in 1902 stated, "Powdered medicine *(umsizi)* belongs to great men [read healers and chiefs]. Ordinary people did not have *umsizi*."[88] Colenso's 1905 dictionary defined umkhando as a "heavy substance, (stone or perhaps metal), used by *izinyanga* in fortifying a chief, weighing down his adversaries."[89] While this may signify that the umkhandos of today did not enter the local African pharmacopoeia at this early date, the naming of powders after such historically powerful muthi is testimony to the value placed on them.

Indian herbs and spices in the African pharmacopoeia include asafetida, ginger, turmeric, cinnamon, cumin, fennel seeds, cloves, and nutmeg.[90] Other medicinal herbs from India that were incorporated into the local pharmacopoeia include bitter oils from the syringa tree, alum, senna leaf, *tulsi* (holy basil), tamarind seeds, fennel, croton seeds, betel nut, cinnabar, frankincense, and myrrh.[91] These various Indian substances generally acquired unique Zulu names, though some words showed linguistic borrowing, often when two medicines are being used for similar purposes—such as "tulsi," which is the same in Zulu and Hindi and is used by both populations for fevers. Others were used to prevent or cure specific African ailments, such as those caused by witchcraft.[92] In addition, Africans borrowed the Indian preparation of aloe—burning it and crushing it into a fine powder—but recognized its origins in its name, *isinhlumbo samandiya* (of the Indians).[93]

While Indian medicinal herbs may have been adopted due to their perceived efficacy in combating certain maladies, they may also have gained added caché as the local medical culture placed greater value on herbs and substances from distant places than on more accessible local herbs. This preference, however, did not always extend to European remedies. The sharing of medical cultures between Indians and Africans thus demonstrates a different relationship than that between Africans and whites. One rural inyanga testified to the acceptance of (male) Indians into African therapeutics by arguing that while he remained incredulous about the reliability and skill of African female inyangas, he had no hesitations about visiting an Indian inyanga and in fact would want to learn from *him*.[94]

POLICING INTERCULTURAL AND MEDICAL BOUNDARIES

Evidence that Africans and Indians crossed cultural boundaries can be seen not only in the rise of Indian inyangas and muthi shops, but in visits that Africans and Indians made to healers and cultural/healing events that were clearly outside their own cultural milieu. Non-Muslim Africans, for instance, visited the Badsha Peer Sufi Shrine in downtown Durban for blessings, and Zanzibaris were hired by but also participated in the yearly festival put on by the Soofie Saheb community.[95] In 1948, an Indian man posing as the reincarnation of Krishna in Verulam claimed "to cure people of any illness, including TB, asthma, cripples, the blind and mute." In response South Africans of all races flocked to see him for healing before he was exposed as a fraud.[96] There is also limited evidence from Kuper's study that finds Indians visiting African healers, particularly better-known Zanzibari ones, as well as direct observations of at least two Africans who visited or participated in Hindu healing ceremonies.[97] Interactions between these two groups can also be seen in the fact that some Indians interviewed by Kuper's group sometimes used the Zulu word "thakathi" to refer to witchcraft, whereas another compared and explained that an Indian spiritual healer called a *wohja* "can work like an inyanga."[98] While patients may have crossed "cultural boundaries," not all approved and some sought to clarify the bounds of culture. For instance, a Mrs. Percieval, a former indentured servant, claimed she had "no belief in witchcraft, they are good for natives only."[99] While an African healer more recently claimed, "Indians don't know anything about this traditional medicines and secondly the people that are buying this medicine are old people so the Indians are getting money, taking our bread from the old people."[100] In each instance the speaker used the Other as a foil to portray him- or herself as nonsuperstitious or as an authentic healer—while simultaneously drawing and reconstructing racial and cultural boundaries. Another example of this can be seen below in one case recorded by Hilda Kuper's assistant in 1953.

In the late 1940s, an Indian man's six-year-old daughter became afflicted by spirits, lost the ability to walk, and talked nonsensically. Desperate, the family brought their daughter to a number of different Indian healers and Hindu temples over a six-year period. At the urging of an African family living in the same yard, the family took their child to a sangoma. According to her father, "A Native witchdoctor [who was] consulted . . . began to speak to a spirit" within the child, and to his surprise, the "child begins to speak in Zulu to him. [The] Doctor puts muthi on her and she sleeps. When she gets up she says something (people) [are] fighting inside her, 5 in all. 3 remained and 2 left. He burnt powder and makes [the] girl inhale smoke uttering Zulu words. 2nd day one went away, then next day another, then 3rd day all. At the end, all [spirits] were gone but [the] girl cannot speak. He tried to get his own spirit to talk to her but could not." Six months later the girl reputedly recovered. After a seemingly successful treatment by the sangoma, however, the girl suffered a relapse, at which time her parents brought her to two Zanzibari healers and another African healer.[101] Last they attended a poojari who admonished the family for bringing the child to "unclean" people and who, upon blowing ash on the girl, instantly cured her. Kuper concludes: "The poojari said that three evil spirits had got into the girl, one of them being [u]fufunyane, an African spirit."[102]

Though this example may slightly exceed our period of study and is probably not typical, it nevertheless demonstrates several important points that can be applied to the earlier period, which shared similar circumstances. First, the family's medical decisions represent Indians' multiple healing strategies and openness to trying various medical practitioners within this medically pluralistic society. Second, the diagnosis of ufufunyane further problematizes the ideas of "culture-bound" illness as it is clear that Indians were being afflicted by some of the same spirits as their Zulu neighbors. What is particularly interesting about this case is that it is an Indian afflicted by African notions of Indian and white spirits.[103] Third, it demonstrates that Indians turned to and used African therapeutics perhaps because they recognized the ailment as being an African ailment or because they did not usually limit their medical choices to one type of practitioner. Fourth, the poojari recognized the ailment as an African ailment, showing that Indian spiritual healers themselves had been influenced by the local ecology of African illness. Last, the healer's comment on cleanliness—a veiled reference to religious purity—is emblematic of someone seeking to patrol cultural boundaries. This may be because African healers represented economic competition, particularly Zanzibari healers who served the same clientele.[104] Or it may signal a threat to the religious authority of the poojari and his sphere of influence.

Competition often occurred between various medical practitioners, and some healers appealed to the government to secure commercial advantage for themselves. Licensed inyangas thus complained of "unqualified" African and Indian healers practicing under a general dealer's license.[105] Likewise certain elite African healers sought to preserve the profession for "genuine *native* nyanga."[106] Like their white medical counterparts, these healers veiled concerns about competition in the language of race and, in 1948, separate development. B. Sikakane, an unlicensed inyanga and member of the Natal Native Medical Association, complained to the chief native commissioner of Pietermaritzburg that Simone Camane, a licensed inyanga who operated in close proximity to Sikakane's practice, was of mixed Indian and African descent. Consequently, he argued, Camane should not be allowed to practice. The chief native commissioner argued that despite Camane's mixed heritage, he spoke only Zulu and had married an African and thus by government standards was considered a "native" and entitled to his inyanga license.[107]

Most policing of local medical boundaries, however, was done not by Africans or Indians but by whites. Although the government had refused to issue inyanga licenses to non-Africans, a practice that followed Shepstonian ideas of customary law, its main concern seemed to be that of reinforcing boundaries between whites and nonwhites. Specifically, it sought to ensure that nonwhite medical practitioners did not attend white patients or sell herbs and remedies that could be confused with those of white chemists. As discussed in the last chapter, the biomedical community had successfully lobbied the government to use the law as a means to distinguish and protect biomedicine from other medical systems. They argued that biomedicine, deemed modern and scientific, was good for everyone regardless of color, but that other medical systems, born of irrationality and superstition, posed both physical and social danger—especially when practitioners and patients crossed the color line. In effect, white patrolling of medical boundaries contributed to the polycultural development of African therapeutics and helps explain why Indian practitioners and substances—more than white or biomedical ones—were adopted.

The incorporation of Indian herbs and substances into African therapeutics correlates to their availability to the Indian community itself. Whereas the government refused to license Ayurvedic or Islamic medical practitioners, it generally did not interfere with Indian home or store-bought remedies. Indians thus maintained access to a wide variety of Indian herbs and remedies, as well as Ayurvedic medicines and even a medicinal holy ash. These were either grown in home gardens[108] or obtained at local dispensaries and Indian shops. In the Durban area, a Sufi dispensary (*dawakhana*) established in the late 1890s in Riverside (a Durban suburb) by Soofie Saheb distributed free medicine for common ailments to the general public on Thursdays. Indians of all faiths and backgrounds attended this free dispensary.[109] A Dr. Cawston noted in 1933 that Indian shops "stock a large assortment of drugs which are used in India."[110] Likewise in 1937 the Natal Pharmaceutical Society noted that the "Burma House" sold patent medicines as well as "Indian remedies."[111] Kuper mentions similar shops along Durban's Victoria Street and Natal's smaller coastal towns that sold Ayurvedic and proprietary medicines in the 1940s and '50s.[112]

Dr. Cawston's initial attention to these Indian shops, however, seems to have stemmed from his observation that they were "patronized even by Europeans." Likewise, the Natal Pharmaceutical Society demanded, unsuccessfully, the license cancellation of the "Burma House" due to the fact that it sold Vicks VapoRub—a patent medicine. White pharmacists feared competition and claimed exclusive rights to sell patent medicine to both white and nonwhite populations. Given the fervor with which the pharmaceutical society pursued African and European shops selling patent medicines during the 1920s and '30s, the lack of complaints about Indian shops by European medical practitioners most likely signifies that Indian shops not only sold predominantly Indian patent medicines but were set up in such a manner that they avoided biomedicine's attention. Indeed Naiker confirms this with her family's own experience: "The one shop in Victoria [street] we were right next to a chemist, Mr. Spates . . . My dad never ever or my grandfather never ever sold anything that they have . . . we never did that."[113] This is important as it means that Indian herbs and Indian patent medicines were able to slide under the radar of biomedical protectionism, avoiding legal interference and thus remaining accessible to both the Indian population and by extension other population groups.

This is in direct contrast to some of the African muthi shops discussed in the previous chapter that looked like white chemist shops and sold patent medicines and consequently were targeted for closure. Biomedical practitioners, fearing the commercial and ideological competition of certain elite African healers, created protective legislation that effectively limited the number of licensed African healers and prohibited African healers from selling Western patent medicines, using preservatives, employing the title "doctor," performing a diagnosis, or caring for white patients.[114] Africans and Indians were not to run muthi or "chemist" shops that could be confused with white chemist shops. Indeed, African and Indian muthi-shop owners I interviewed within the larger metropolitan centers always emphasized that they *dispensed* herbs and did not diagnose or mix herbs, something legally relegated to biomedical doctors and chemists.[115] In this way, the patrolling by and intervention of South Africa's biomedical community forced both African and Indian healers to adopt white biomedical ideas of what constituted "traditional" African or Indian medicine and therapeutics. In essence, "tradition" was defined as the absence of biomedical tools, titles, packaging, and substances—the antithesis of what was white and biomedical.

CONCLUSIONS

The majority of today's muthi shops in KwaZulu-Natal's city and town centers are still Indian owned, though this has recently begun to change. The Indian muthi shop endures for several reasons. First, it offered a resource that was in demand and culturally relevant. Second, as the muthi trade was chased out of the streets and into the shops, Indians had the capital to open stores and the strength of family syndicates to maintain them. Third, the historical legacy of white rule and relative privilege enabled many Indian businesses to thrive during the twentieth century and under apartheid and the Group Areas Act. Last, Indian healers and muthi shops managed to avoid the scrutiny of the biomedical profession by carefully abiding to what biomedicine deemed "traditional" medicines. In a period of rapid urbanization and African alienation from the land, muthi shops served as an important link between town and country and also as a forum for exchanges between Indian and African healers and their patients.

In exchanging medical knowledge, Indians and Africans engaged in cultural transactions that transcended both medical cultures and transformed the way African medicine was and is practiced in Southern Africa. By highlighting the transfer and production of medical traditions and substances between population groups, we can see how African therapeutics are indeed the result of the historical encounters between Indians, Africans, and whites. Likewise we see that delineating some influences and appropriations between population groups has broader applications for demonstrating the nature of that contact, principally in the era before apartheid. This is particularly important in a country that has long ignored and in many cases denied the interactions of its diverse communities.

The need to disentangle some of the polycultural threads of African therapeutics becomes evident when one considers the implications of common misperceptions for public policy. Understandably, today's policymakers'—let alone the general public's—knowledge of Indian-African relations still reflects the success of apartheid and apartheid education. Many South Africans with nominal knowledge of Indian-African cultural mixings assume they are superficial, and in the case of Indian inyangas and muthi shop owners, purely commercial. Unfortunately, some policymakers who work with African healers assume incorrectly that muthi-shop owners of Indian descent know nothing about African medicine and that it is Africans in the back of the shops who were/are the real holders of that knowledge. The implication is that Indians are exploiting African knowledge for their own commercial gain. Interestingly many African healers who interact with Indian muthi shopkeepers do not hold this opinion.[116] At a time, however, when indigenous knowledge systems are finally being recognized and supported by a new South African government, it is particularly troubling to realize that policymakers drawing up legislation regarding indigenous knowledge systems are unaware of the contributions of Indians to African medicine. It is partly for this reason that it is so important to delve deeper into the nature of African-Indian encounters. This chapter raises important questions regarding the "indigenous" nature of African medicine by demonstrating that Indian inyangas were and are not only the holders of so-called African indigenous medical knowledge but its shapers and contributors as well.

Epilogue

IN AUGUST 2004 South Africa officially recognized its "indigenous" medical systems and began the process to legalize the practice of traditional healers. After years of being criminalized under white minority rule and largely condemned by the biomedical community, South Africa's 350,000 traditional healers must soon obtain a government license to practice. Healers will be restricted from treating fatal diseases such as cancer and HIV/AIDS, but given their popularity—80 percent of the South African population consult such healers—they are seen by many as a necessary component of healthcare delivery and as important agents for educating the populace on the realities, perils, and prevention of HIV and AIDS. Incorporating traditional healers and medicine into a state medical system that has overwhelmingly preferred the biomedical sciences has not been easy. As this book has shown, these two medical cultures not only embrace different ideas about the body and the origins of illness, but share a history of commercial and ideological competition as well as different relations to state power. The legalization of traditional medicine thus presents a number of difficult questions. One of the most important is: How will a largely local, unsystematized, nonhierarchical, and oral collection of therapies that as of yet is largely unregulated come to be systematized and brought under the regulatory eye of government? Given that governments tend to favor institutional sciences and those who are Western educated and bureaucratically literate, is it possible to rectify the imbalances of power and enable healers to be meaningfully incorporated into a state-sanctioned system of medical pluralism? Clearly South Africa is not the first country to undergo such a transition;[1] nevertheless this is an important process to watch.

In addition to the challenges of regulating African medicine, healers in the future will have to cope with the growing HIV/AIDS epidemic,[2] overharvesting of medicinal plants, protecting intellectual property rights, avoiding exploitation for commercial gains, and possibly losing control over the ways in which future generations utilize traditional medicines. Changes in South Africa during the 1990s profoundly altered the stakes of traditional medicine and created new stakeholders that now include pharmaceutical companies, government departments, and university research labs. Bioprospecting of indigenous medicinal plants is now seen as a legitimate and necessary government project. The health minister herself has stated, "South Africa is blessed with a rich heritage of medicinal plants that through sustained research and development, could offer a solution to some of the common health problems the world is grappling with."[3] An integral part of this project is collecting and collating existing local medical knowledge in a form that is useful and knowable to government, university labs, and pharmaceutical companies. While the collection of medicinal botanical knowledge has roots in the colonial period and the collaboration of healers and scientists is not new, the University of Cape Town's Traditional Medicine Program began a new type of collaboration in 1994. This project brought in traditional healers during its initial phases, helped them to write articles, produced a "Traditional Healer's Primary Care Book," promised benefit sharing in the case of successful bioprospecting, and has successfully created a national databank of South Africa's medicinal plants. The creation of a database has important implications in its ability to transform one type of knowledge and power (folk or "traditional") into another (institutional science). Such transfers of knowledge could have important repercussions to both knowledge communities. How has this partnership been cultivated, and what do the various stakeholders perceive as the benefits of such interaction? These are important questions for future research and ones that can potentially help other medically plural societies.

A number of factors in the 1990s changed government and scientists' ideas about the importance of African healers and medicinal plants. First, the HIV/AIDS crisis that exploded in the 1990s opened up pressures for new collaborative efforts between healers and biomedical practitioners. While such schemes had a number of detractors, on both sides, others could point to a number of successful collaborative projects.[4] More importantly the end of apartheid bought a flush of international funding, with donor agencies seeking to support collaboration and HIV/AIDs training for healers. Second, the "legitimacy" of African medicines gained a boost on a national level when Thabo Mbeki assumed the presidency in 1998 and articulated his vision of an "African Renaissance." An important component of which was "Indigenous Knowledge Systems" or IKS—the art, science, technology, practices, and knowledge systems passed down through South Africa's generations. While the notion of IKS was not particularly new, its support on a government level was a radical break from the apartheid government, which had openly disparaged and ignored African medicine and science.[5] Third, the organization of traditional healers seeking legalization and licensing of their profession affected the ways that both biomedicine and healers saw the role of traditional healers and medicine in South Africa's future healthcare delivery. Fourth, the financial success of the medicinal herb *hoodia* held out hope for both the South African government and its pharmacology labs, who coincidently (or not) have been vigorously touting the benefits of IKS ever since.[6] While Medical Research Council funding for the Traditional Medicines Center at University of Cape Town seems to have come a bit earlier than this, recognition of the commercial potential of South Africa's indigenous cultures has greatly increased government interest, investment, and support. Finally, long-standing environmental concerns moved botanists, government leaders, and healers to consider new options with regard to the harvesting and processing of traditional medicinal plants.

BIOPROSPECTING TRADITIONAL MEDICINES: THE CASE OF HOODIA

In 1996, South African government scientists in the Council for Scientific and Industrial Research (CSIR) isolated and patented compound P57 from the Kalahari's *Hoodia gordonii* plant—an appetite suppressant used by the San people in southern Africa to sustain them during long hunts or during times of food scarcity. In 1997 they licensed its patent to the British pharmaceutical company Phytopharm for twenty-one million dollars. Phytopharm in turn subleased it for commercial development to U.S.-based Pfizer Corporation and then to Unilever, the maker of Slimfast.[7] What is unique and interesting about the hoodia case was the decision by the CSIR to participate in benefit sharing with San communities. Negotiations were prompted when a Pfizer representative in the UK remarked that the San—the original discoverers of this increasingly popular herb—were extinct.

Legally there is little precedent under international law to support indigenous peoples' claims to patent traditional medicine. Existing international patent law as spelled out in the World Trade Organization's Agreement on Trade-Related Aspects of Intellectual Property Rights (TRIPS) (1986–94, Uruguay Round) prohibits the patenting of whole plants. Instead patents require new innovations and discoveries. Thus new plant varieties, such as Monsanto's bioengineered crops that resist the weed-killer "Round Up" or discoveries like compound P57 can be patented, whereas the hoodia plant cannot. An integral part of patenting and bioprospecting involves subjecting plants to laboratory testing with the aim of isolating active compounds. Efforts are then generally taken to synthetically reproduce such compounds, both to reduce the costs of production but also to create greater chemical stability and quality control. Dependence on laboratory testing thus makes it very difficult for "traditional" forms of knowledge to be patented. As prescribed, international patent law cannot recognize the originators who discovered the use (medicinal or otherwise) of a plant. The inequities of this system, which privilege those with capital and laboratories over those who are knowledge rich but "research" poor, has led a number of researchers to call this process "biopiracy."[8] While moral imperatives for benefit sharing may be missing from international patent law, local patent laws can provide some protection. Yet even these laws presume that patented knowledge is exclusive — to either an individual or a corporate author. This thus raises questions of how to compensate a community such as the San or a smaller number of specialists such as traditional healers for indigenous knowledge that contributes to the discovery of active compounds like P57. Furthermore, how do such communities even afford the expense of securing an initial patent, or the yearly costs of license renewal, both of which may be beyond their reach? The Convention on Biological Diversity (1992) may provide some protections, given that signatories must ensure "the conservation of biological diversity, the sustainable use of its components and the *fair and equitable sharing* of the benefits arising out of the utilization of genetic resources."[9] While such language might hint at benefit sharing, it is not prescriptive and thus leaves open the question of whom to compensate (nations, communities, or healers) and how.

In the case of *hoodia*, the San as a community have already received R260,000 from CSIR, and future royalties have been projected as high as R8 to 12 million. In addition, the San not only were promised a percentage of future royalties, but signed an agreement to ensure their participation in growing and cultivating the plant.[10] This decision may in fact be the most lucrative, given the world's desire for *hoodia* and the difficultly in profitably synthesizing P57. Such difficulties in fact led Pfizer to drop the project in 2003.[11] This did not dissuade Unilever, however, which paid £6.5 million for exclusive rights to the compound and is conducting further trials. The company is expected to set up large *hoodia* farms in South Africa and begin producing products by 2008.[12]

While P57 has thus far remained undeveloped,[13] the sale of the patent and a preliminary unpublished study by Phytopharm generated much attention and prompted the sale of *hoodia* and imitation *hoodia* by natural-remedy companies in South Africa and abroad. The increased demand for *hoodia* led to its being placed on the endangered species list, and its export out of South Africa is now strictly controlled by the South African government.[14] Nonetheless this has not stopped poachers from attempting to steal the plant, nor natural-supplement companies from claiming to sell it.[15]

The *hoodia* example shows us that the process by which traditional medicines are currently being absorbed into the biomedical market is both similar to and different from such processes in the past. Primarily, we see that the process of bioprospecting means extracting African medicinal plants from both their physical and their social environment, with little acknowledgement of (let along compensation for) the persons who initially utilized them. In the past any acknowledgment that linked African medicinal plants to Africans was generally disparaging. Indeed it was the process of scientific inquiry in which chemists and pharmacologists extracted African plants that made African *muthi* (medicine), formerly cloaked as superstitious and unscientific, both knowable and valued to the scientific world.[16] While *hoodia* initially followed this same scenario, changes in South Africa, including its signing of the Convention on Biological Diversity in 1995,[17] along with some strong legal coercion on the part of San lawyers made it impossible to completely disassociate the San from *hoodia*.[18] Second, bioprospecting of African medicines, like other traditional medicines around the globe, generally has involved chemical testing of plants for their commercial value and eventual use by non-indigenous populations. Clearly this is the case with *hoodia*, and it is hard to escape the much-remarked-on irony that the San turned to *hoodia* because they lacked food, whereas obese Americans may use it due to an overabundance of food. Finally, it is difficult to determine what impact *hoodia*'s commercial development will have on San communities. While the San have enjoyed greater recognition around the world, with the likes of *60 Minutes*' Leslie Stahl stomping through the Kalahari, what will such attention and monies bring to these communities? As authors Posey and Dutfield ask, will the tenor of relations between local communities and commercial enterprises be one that shows respect for local notions of the environment, or will it create a cycle of dependency?[19]

HEALING THE BODY AND THE NATION IN THE TIME OF HIV/AIDS

The rate of HIV/AIDS in South Africa exploded in the 1990s, and already it is estimated that 1.8 million South Africans have died of the disease. By mid-2006 approximately 5.5 million, or 11 percent of South Africa's total population, was HIV-positive. The highest rates were seen in women between the ages of 25 and 29—33 percent—and in men ages 30 to 34—27 percent.[20] Such high rates of infection reflect multiple factors. In part they are due to South Africa's high rate of sexually transmitted diseases and infections, which ease transmission of the HIV virus, and other major diseases such as tuberculosis and malaria, which compromise the immune system. Other less tangible factors include a long history of inequality, migrant labor, landlessness, poverty, and overcrowded and unhealthy living conditions that have helped break down the family and suppress individual immune systems. Furthermore the declining power of women that accompanied white rule resulted in limited reproductive choices for African women and their ability to demand that partners practice safe sex. A large dose of government inaction during the early period of the epidemic meant missing the less expensive and easier task of prevention, which could have stemmed the epidemic. This has been compounded by the fact that a post-apartheid South Africa inherited a rather uneven and poor biomedical infrastructure, and that the cost of antiretrovirals and other drugs used to treat opportunistic infections associated with AIDS has historically been very high.[21]

Africans' reactions to the disease also tended to be colored by a long and contentious history between "the West" and Africa. Shortly after winning the 2004 Nobel Peace Prize, winner Wangari Maathai, the Kenyan environmentalist, shocked much of the western world by suggesting that the West had crafted HIV as a biological weapon to be released on Africa. Such beliefs, however, are largely accepted in many parts of Africa, where the acronym AIDS has been dubbed the "American Invention to Discourage Sex" and the French acronym SIDA is translated equally cynically as "*Syndrome Inventé pour Décourager les Amoureux*" (Imaginary Syndrome for Discouraging Lovers). Such skepticism comes from memories of moralizing missionaries and colonists, who not only introduced the "missionary position" but condemned polygamy, premarital sex-play, and sex education. Furthermore much western scientific and nonscientific discourse surrounding AIDS has blamed Africans for the epidemic — both in terms of their behavior and as the originators of the virus. Only in late 2006 did the medical journal *Lancet* publish a study proving that Africans were no more sexually promiscuous than other population groups — a question long pondered by researchers who sought to explain why Africans had higher rates of the disease than other population groups.[22]

Given their history, it would not be surprising if black South Africans harbored a degree of cynicism toward biomedicine, particularly as a result of apartheid's coercive and racist Population Control Program. Concern over reducing the black population led the apartheid government to make fertility clinics widely available and to offer free sterilizations. While compulsory sterilization never became government policy, many white doctors acted on their own accord by inserting IUDs and performing tubal ligations, and the private sector sometimes used coercive measures to ensure that black women remained infertile. Biomedical doctors were on the frontlines of this effort, and thus many South Africans remain skeptical of anything that can be interpreted as fertility control.[23] Likewise the apartheid government's claims that anti-apartheid activists had AIDS made the new government and the African populous initially wary of the former government's predictions about the disease.[24] The inaction and silence of South Africa's first democratically elected government was followed by President Thabo Mbeki's dismissal of biomedical explanations of and remedies for HIV/AIDS and the decision of multinational pharmaceutical firms to sue South Africa in the late 1990s to prevent it from purchasing more affordable generic HIV/AIDS drugs. Even though this pharmaceutical suit was eventually withdrawn, these last acts only exacerbated public skepticism of biomedicine.[25]

Partly in response to these historical pressures and the high price of antiretroviral drugs, Anthony Butler argues that Mandela and Mbeki's offices adopted a "nationalist/amelorative" paradigm to resolve the AIDS crisis. Instead of focusing on a "biomedical mobilization plan," for which antiretrovirals would have played an important role, the government instead focused on reducing poverty, offering palliative care, and encouraging better nutrition and the use of traditional healers.[26] The health minister earned the unflattering moniker "Dr. *Beetroot* Msimang Tshabalala" (1999–present) because she emphasized the importance of good nutrition in lieu of antiretrovirals as a defense against AIDS. She also promoted use of the African potato *(Hypoxis hemerocallidea)*—a traditional medicine shown to boost the immune system.[27] With regard to healing the body and nation in the time of HIV/AIDS, healers played a largely informal role rather than a formal one initiated by the government. African healers often offered hope and comfort to HIV/AIDS patients, particularly when biomedicine turned away many unable to afford the expensive regimen of pharmaceutics associated with HIV and AIDS. Several healers whom I interviewed back in 1998, when stigma was a much larger issue, told me that they housed or provided hospice care to people they knew or suspected of having HIV/AIDS. This is particularly important given the hostile climate that left (and can leave) the afflicted feeling ostracized or hopeless.

The practical aspects of adding healers to the public health delivery system eventually became clear to many local governments and nongovernmental organizations. By the late 1990s, a number of healers had participated in HIV/AIDS workshops. These aimed not only to educate healers on the symptoms of AIDS, but to encourage healthful healing practices such as sterilizing or not reusing razor blades, needle-like devices, and enema horns. Healers were also trained so they could disseminate information to prevent the transmission of HIV. Furthermore these healers counseled patients on nutrition and seemed to have had some success in their treatment of secondary infections, which also helped increase the morale and well-being of patients.[28]

Between 1999 and 2000 South Africa's Medical Research Council conducted a study regarding the acceptability and effectiveness of traditional healers as supervisors of tuberculosis treatment. As a result of the study they found that healers provided a number of advantages for care, including accessibility, shorter waits, good patient compliance, and overall care of patients' well-being. This led the MRC to recommend that healthcare authorities "consider integrating traditional healers into other aspects of healthcare including voluntary counseling and testing for HIV and for home-based care for people with AIDS."[29] Healers were included as one out of fifteen civil society representatives on the National AIDS Council.[30] On a regional level, 2003 marked the signing of a "memorandum of understanding between the Indigenous Healers of KwaZulu-Natal and the Nelson R. Mandela School of Medicine." At this signing the KwaZulu-Natal Minister of Health reiterated the need to incorporate healers into the fight against AIDS: "We know that there are many traditional medicines that have tremendous effect in alleviating opportunistic infections and boosting the immune system. We are soon going to be bringing together traditional health practitioners to share with us their experiences so that we discuss some way of incorporating these into our response to HIV and AIDS."[31] More recently this same medical school began working with the local traditional Health Practitioners Councils to train 375 healers to test patients for HIV, keep records of their progress, and refer patients to AIDS clinics where they can receive antiretroviral drugs.[32] Other groups, such as the Treatment Action Campaign (TAC), an advocacy group for affordable medicines for HIV/AIDS patients, have likewise embraced the role that healers can play in fighting the epidemic.[33] TAC not only focused an entire newsletter on the subject of traditional healers and public health, but began its own four-day workshops for healers in 2005.[34]

While the history of interaction between South African healers and biomedical practitioners has been contentious, clearly the importance of such interaction has gained increasing recognition, particularly in light of the current HIV/AIDS epidemic, which has reached levels beyond the capacity of biomedical care. Lately new studies have shown that two herbal immune boosters, African potato and *Sutherlandia*, significantly reduce the effectiveness of antiretroviral drugs and "may put patients at risk of treatment failure, viral resistance or drug toxicity."[35] Clearly it is necessary that medical practitioners—traditional or biomedical—know if their patients are on antiretrovirals or herbal remedies so they can avoid herb-drug interaction, and, in the case of healers, avoid prescribing emetics and enemas, which make antiretrovirals ineffective and dehydrate patients.[36] Unfortunately some healers have continued to encourage unsafe sexual and medicinal practices to the detriment of their patients. This is largely because biomedicine has focused on sexual control and ignores the issues of fertility/childbearing, which generally confer important social status to African men and women.[37] Some healers thus erroneously offer "blood cleansers" that allegedly allow a man to temporarily cleanse the body and impregnate his partner without passing along the disease.

The incorporation of various types of healers and African medicine into South Africa's healthcare system, however, raises a whole host of issues. Along with this incorporation are some major issues that South Africa will need to face, regarding both healthcare delivery, issues of intellectual property, and sustainable harvesting of medicinal herbs. These challenges have prompted new thinking about and rethinking of how healers should deliver and regulate healthcare, as well as a new means of establishing their medical authority.

ENVIRONMENT

Longstanding environmental concerns regarding biodiversity gained urgency as the lifting of government restrictions on population movements and relaxed enforcement of informal trading resulted in a resurgence of the muthi markets. The increased mobility of the entire African population made the muthi trade even more lucrative and drastically accelerated the collection, exploitation, and disappearance of certain medicinal plants. The problem of overcultivation, however, is not new. It can be seen as early as 1898, when botanist Medley Wood mentioned the extinction of umondi (Mondia whitea) in the Durban area, reflecting the pressures of increased early urbanization. The eradication of indigenous forests and vegetation by expanding sugar estates and tree farms in the nineteenth and twentieth centuries further reduced the habitat of indigenous species, clearly contributing to decreased supplies of local herbs, roots, and barks, and no doubt contributing to overcultivation in the remaining areas. Consequently, medicinal plants, particularly slower-growing roots and trees, became increasingly rare as the muthi trade expanded. White botanists began commenting on Natal's loss of local medicinal botanicals in the 1930s and '40s.[38] These types of muthi not only became more expensive as they became increasingly rare but had to be gathered from farther and farther afield. Likewise healers from the rural areas found they now had to come to town to buy certain herbs they had once gathered for free. While many healers boast of harvesting medicinal plants only in a sustainable fashion, usually on a case-by-case basis in the rural areas, the muthi trade and muthi shops historically prompted nonexperts to collect different herbs, barks, roots, animal skins, bones, and fats for sale in towns and cities. By the 1940s, gathering seems to have become an increasingly female occupation, as rural women brought herbs to town to sell to traders, herbalists, and muthi shops.[39] During the '40s as during the 1990s, gatherers were generally poor and unskilled. Few invested in sustainable harvesting, and many participated in overharvesting because they were reluctant to leave behind muthi that could be collected by competitors. Liberalization of the muthi markets since the end of apartheid has only increased the environmental impact of such collecting methods. Durban's major muthi market at Warwick Junction, which derives its origins from the same trade that began at the beginning of the twentieth century, was said to be trading over 490 tons of muthi a year by the late 1990s.[40]

Environmentalists warned that the muthi trade, in conjunction with invasive plant species and other land and population pressures, would lead to the extinction of some of South Africa's important flora. Such concerns led botanists in the early 1990s to begin a rigorous campaign to start muthi gardens to encourage healers to cultivate rather than wild harvest specific endangered medicinal plants. The idea was to decrease demand and prices in the hopes of decreasing the harvesting of these plants.[41] Such gardens and nurseries have helped to link gatherers, healers, conservationists, educators, and government administrators.[42] Also the farming of medicinal plants might create job growth in the rural areas. In addition to conserving biodiversity, there were apprehensions about what the loss of botanicals may mean for future medical cures. Environmental and medical concerns alike have played an important role in compelling both scientists and government leaders to support a new type of bioprospecting that conserves useful plants. These in turn led to government support for a national database, TRAMED III, that lists the three hundred most used traditional medicines. It was also part of the reason that pharmaceutical companies such as Hoechst/Noristan began working on a database of South Africa's medicinal plants in 1974.

Furthermore botanists have been trying to get healers to consider new means of processing herbs. In their rough and unprocessed form, much is likely to be wasted as plants loose potency over time and others are lost due to mold and fungus, leading them to be discarded. Processing herbs by grinding them into powder would presumably mean they would keep longer, producing less waste and thus reducing the price of muthi. The question is whether or not healers and patients can be convinced to administer and take medicinal herbs in this new form. Testing and training would also be required to ensure that the proper amount of herbs are administered.

As this book has shown, African healers have been seeking to convince governments of their legitimacy since the early 1930s. To do so healers organized themselves and made appeals to government administrators in the language and symbols of science and modernism. One healing association from this early period that sought government recognition even referenced botanists Watt and Breyer-Brandwijk's work on medicinal plants (1932). Surely if the scientific labs of white South Africa had tested and recognized the power of traditional medicines, how could they have prevented Africans from using them? This need to situate African medicines within the framework of "Western" or academic knowledge was not merely rhetorical. In fact it is one of the driving forces behind the current collaborative efforts of healers, academic researchers, government departments, and pharmaceuticals, though many healers have maintained an understandable degree of skepticism of this approach.

Both sentiments were reflected by Muzi Mthemjwa, a Durban isangoma and artist, in 1998:

> Now that the government has changed, they are giving people [healers] the opportunity to change themselves and improve themselves. Now for instance I just buy a grinding machine instead of wasting material like I do when chopping like this . . . Those who are becoming *sangomas* should go to school to learn mathematics to learn how to mix what with what. Our laboratories are our ancestors, they say "no you've made too much muthi, just take some of that out, put more of this stuff and this stuff." But now we cannot rely on our ancestors to prove ourselves, we must have laboratories. We must have people who can purify and test our medicines to see if it is good for selling to other countries. We must have our own laboratories and purifiers to keep pharmaceuticals from taking advantage of us.[43]

Mthemjwa gave this answer in response to my question of whether or not "Zulu medicine" could "modernize." While this answer may reflect Mr. Mthemjwa's assumptions about me and my background, it also highlights a certain tension that existed in 1998 and continues today—that is between modernizing African medicine and legitimizing its existence through its roots to a precolonial past. Despite the recognition that a healer's medical authority comes from "traditional" or "indigenous" sources such as one's connection to ancestors, the language of science and modernity is used as a metaphor to describe the accuracy of the ancestors. Mthemjwa's comments about the necessity of including mathematics, as well as purifying and testing African medicines in a laboratory point to a need to establish "traditional" healers' medical authority under a biomedical as well as a local medical model. Mthemjwa's request for a lab seems to signify cutting out a new type of space, both figuratively and literally.

Healers' enthusiasm for modernizing African medicine and legitimating its efficacy within a biomedical model reflects healers' responses to global and local conditions, such as a booming natural-remedy market, bioprospectors from international pharmaceuticals, a government supportive of Indigenous Knowledge Systems, and pending government legalization of their profession. Current healers, like many of their urban predecessors in the 1930s, are thus trying to figure out ways of balancing what is "traditional" about African medicine while at the same time using biomedical ideas of the body, research, and professionalization to create advantages for themselves.

The government's legal recognition of healers emerged from a largely healer driven process. Up through the 1980s, a handful of inyangas in Natal and Zululand had been the only legally licensed healers in the country. While healers had been actively organizing into regional and national healing associations since the 1930s, the 1980s saw greater efforts to nationalize these organizations.[44] Such movements were motivated in part by the need to combat what healers saw as an increase in "charlatanism." Associations sought to counteract this by increasing standards of professionalization for their members and vigorously lobbying local and regional governments for legal recognition. With the change in government in 1994, such efforts only intensified. Licensing not only would help healers to better implement quality control over who could practice as a legitimate healer, but represented government validation of their occupation. Such validation meant tangible monetary benefits, including reimbursement for medical aid, the ability to bring deadbeat clients to court, and the chance to gain government monies and support for healer initiatives. The hope of government licensing served as a backdrop to many of the collaborative projects that healers engaged in with government, biomedical practitioners, and academic researchers in the 1990s.

In some cases these collaborative efforts showed immediate benefits to both parties. In the case of Durban, healers' interactions with local administrators led to gains on the national level with regard to legal recognition. For instance, in 1995, Mr. V. T. Mkize, head of Environmental Health, a subdivision of Durban's Department of City Health, set up a body composed of department members and heads of local healing associations in order to assess local healers' needs. Healers complained about the Warwick Junction muthi market and the ways in which herbs were marketed in both an unsanitary manner and without regard to local customs. Many of the women collectors who brought muthi to the city violated taboos by sleeping at the market to protect their merchandise. From the city's perspective the muthi market also posed health and safety risks. As a result of numerous discussions, a new partially covered market that enabled healers to secure their merchandise at night was built near Russell and Victoria Streets. The negotiations over this market did not proceed quite as smoothly as Mkize depicted, but concrete changes were visible by 1998.[45] Thus unlike the early 1990s, when healers and collectors had to worry about being raided by the police or the Natal Parks Board (concerned about the illicit gathering of endangered plants and animals), the late 1990s transformed the government into a viable ally. Likewise, the umbrella organization initiated and mentored in part by Mkize trumped other regional healing associations in terms of organization and knowledge-testing of healers, priming KwaZulu-Natal to becoming the first province to begin licensing healers.[46]

This reinventing of traditional medicine—through legalization and collaboration with biomedicine, however, has not gone uncontested. As mentioned earlier, the KwaZulu-Natal Traditional Healing Council has continued to debate the legitimacy of processing muthi. Likewise many individual healers express skepticism about traditional healing councils and assume their leadership is more interested in money and power than in public health. Others argue that licensing will only increase healers' and therefore patients' expenses while adding on the burdens of "professionalization" and state regulation.

Furthermore, patients have not always embraced healers who have advocated cultural change and attempted to expand traditional medicine in new ways. Isangoma Queen Ntuli learned this the painful way. She had shared her home/office in the township of Umlazi with a biomedical doctor with whom she did cross-referrals but had to close down due to lack of patients. Ntuli, who studied homeopathic remedies as well as African medicine, used both in her practice. Her consultation room in 1998, located on the outskirts of Durban's city center, was in a small room on the first floor of an apartment complex that looked onto a center courtyard. The only sign that hung in her window advertised her consultation room hours—8 am to 5 pm Monday through Friday—and listed ailments for which she could provide remedies. Ironically, patients who sought nonbiomedical care did not have problems using biomedical categories, such as B.P., diabetes, and arthritis, which were some of Ntuli's specialties.[47] Ntuli was involved in the creation of Freedom Hospital (a hospital of African healers) in Durban, which opened in July 1996. This hospital trained "traditional nurses" who in addition to their work at Freedom were required also to obtain a first-aid certificate. Diagnosis at the hospital included local healing forms such as bone throwing and inhaling impepo, but could also include the drawing of blood, which was then sent to a nearby medical facility for analysis.[48] The hospital closed down in 1998 because, as Ntuli said, "these things were too new." There were also rumors about financial mismanagement and the creation of false expectations that nurses' training guaranteed employment at the hospital. There was some resentment when the hospital could not hire all the trained "traditional nurses" who, with a previously unrecognized occupation, found it difficult to obtain employment. Nursing was commonly seen as the role of the healer, her assistants, or of the female family members of the patient. Given that Ntuli had seen both her collaborative clinic in Umlazi and Freedom Hospital close down, she seemed to think at that moment that Africans were not ready for this type of collaboration between African and biomedical therapeutics. While patients may not be ready for such "reinventions" of traditional medicine, many healers such as Ntuli, who had become the secretary general of KwaZulu-Natal's Traditional Healing Council by 2002, saw collaboration with government and biomedicine as an important and viable strategy for South Africa's future traditional healers.

Notes

INTRODUCTION

1. "To Chief Commissioner of Native Affairs Department from Q. Cele," received September 8, 1934, Pietermaritzburg Archive Repository (hereafter PAR), CNC, 50A.

2. Rex v. Ngcobo, 1941 *South African Law Review*, 413; and court case *Rex v. Mafavuke Ngcobo*, South African National Archives Repository (hereafter SAB), GES, 1788, 25/30M.

3. The term *biomedicine* refers to what has otherwise been termed allopathy, Western medicine, or cosmopolitan medicine. Given that this medical system has developed and is practiced throughout the world, I prefer the term biomedicine, which highlights its attachment to the body.

4. My emphasis. 1932 Code of Native Law.

5. My emphasis. On appeal, 1941 *South African Law Review*, "Rex v. Ngcobo," 423.

6. "Rex v. Ngcobo," 228.

7. Ibid., testimony of Qhoboshene, 182, testimony of Kuzwayo, 265.

8. Ibid., 40.

9. Ibid., testimony of Kuzwayo, 265.

10. Ibid., 263.

11. Ibid., 37.

12. "Rex v. Ngcobo," 1941 *South African Law Review*, 430.

13. My emphasis. Ibid., 428.

14. Ibid., 413.

15. Ibid., 422–23.

16. This court case gives evidence for both Indian uses and processing of croton seed, and use of senna.

17. Traditional Health Practitioners Bill 2003, Republic of South Africa, B66–2003. Portion extracted: "Communicated from ancestors to descendents or from generations to generations, with or without written documentation, whether supported by science or not." First passed in 2004, but not signed into law until January 2008.

18. The term "medical system" implies a closed and internally logical and integrated medical culture. Murray Last asks whether this notion is actually useful in a medically plural society such as Hausaland, Nigeria. I have likewise avoided the term "system" for Last's more open and useful term of "medical culture." M. Last, "The Importance of Knowing about Not Knowing: Observations from Hausaland," in *The Social Basis of Health and Healing in Africa*, ed. S. Feierman and J. Janzen (Berkeley: University of California Press, 1992), 393–409.

19. E. E. Evans-Pritchard, *Witchcraft, Oracles and Magic among the Azande* (Oxford: Oxford University Press, 1937), 15.

20. J. Vansina, *Oral Tradition as History* (1965; repr., Madison: University of Wisconsin Press, 1985).

21. J. Vansina, *Paths in the Rainforest* (Madison: University of Wisconsin Press, 1990), 258.

22. Ibid., 260.

23. V. Y. Mudimbe, *The Invention of Africa: Gnosis, Philosophy, and the Order of Knowledge* (Bloomington: Indiana University Press, 1988).

24. For a good overview of the historiography of tradition in Africa see T. Ranger, "The Invention of Tradition Revisited: The Case of Colonial Africa," in *Legitimacy and the State in Twentieth-Century Africa*, ed. T. Ranger and O. Vaughn (New York: Palgrave, 1993), 62–111; and A. Spiegel and P. McAllister's introduction to *Tradition and Transition in Southern Africa* (London: Transaction Publishers, 1991), 1–8.

25. N. Kodesh, "Renovating Tradition: The Discourse of Succession in Colonial Buganda," *International Journal of African Historical Studies* 34, no. 3 (2001): 511–41.

26. S. Feierman, *Peasant Intellectuals: Anthropology and History in Tanzania* (Madison: University of Wisconsin Press, 1990).

27. For African mediators see C. Hamilton, *Terrific Majesty: The Powers of Shaka Zulu and the Limits of Historical Invention* (Johannesburg: Witwatersrand University Press, 1998); N. Rose Hunt, *A Colonial Lexicon of Birth Ritual, Medicalization, and Mobility in the Congo* (Durham, NC: Duke University Press, 1999); J. Iliffe, *East African Doctors: A History of the Modern Profession* (Cambridge: Cambridge University Press, 2002); A. Digby and H. Sweet, "Nurses as Cultural Brokers in Twentieth-Century South Africa," in *Plural Medicine, Tradition and Modernity, 1800–2000*, ed. W. Ernst (London: Routledge, 2001), 113–29; and S. Marks, *Divided Sisterhood: Race, Class and Gender in the South African Nursing Profession* (New York: St. Martin's Press, 1994).

28. H. Sibisi (Ngubane), "The Predicament of the Sinister Healer: Some Observations on 'Ritual Murder' and the Professional Role of the Inyanga," in *The Professionalisation of African Medicine*, ed. M. Last and G. Chavunduka (Manchester: Manchester University Press, 1986), 189–204.

29. For a good overview of the ethnical and cultural complications associated with controlling witchcraft as well as the Ralushai Commission Report see J. Hund, ed., *Witchcraft Violence and the Law in South Africa* (Pretoria: Protea Book House, 2003). Also see I. Niehaus, *Witchcraft, Power, and Politics: Exploring the Occult in the South African Lowveld* (Cape Town: David Philip, 2001); A. Ashforth, *Witchcraft, Violence, and Democracy in South Africa* (Chicago: University of Chicago Press, 2005).

30. During my stay in South Africa in 1992, I heard a number of national radio shows in which healers made such claims. Recently traditional healers in KwaZulu-Natal have asked the press for the right of first notification so that they can respond to any articles that might negatively affect their image. A. Devenish, "Negotiating Healing: The Professionalisation of Traditional Healers in KwaZulu-Natal between 1985 and 2003" (master's thesis, University of Natal, Durban, 2003), 77.

31. Recently government designation has changed to refer to healers as "traditional medical practitioners." In this case, "tradition" seems to be acting as a label and branding of local and African medicine. Traditional Health Practitioners Bill 2003.

32. F. Scorgie, "Virginity Testing and the Politics of Sexual Responsibility: Implications for AIDS Intervention," *African Studies* 61, no. 1 (2002): 55–75; S. LeClerc-Madlala, "Virginity Testing: Managing Sexuality in a Maturing HIV/AIDS Epidemic," *Medical Anthropology Quarterly* 15, no. 4 (2001): 533–52.

33. L. Vail, introduction to *The Creation of Tribalism in South Africa* (Berkeley and Los Angeles: University of California Press, 1991), 11–14. E. Shils, *Tradition* (Chicago: University of Chicago Press, 1981).

34. Bioprospecting for environmental reasons, different than bioprospecting or biopiracy for drugs, see W. V. Reid, "Bioprospecting: A Force for Sustainable Development," *Environmental Science and Technology* 27, no. 9 (1993): 1730–32. As compared to the work of V. Shiva *Biopiracy: The Plunder of Nature and Knowledge* (Boston: South End Press: 1997).

35. Phillip Kubukeli, interview July 2002, Cape Town.

36. Devenish, "Negotiating Healing," 85.

37. The Natal Parks Board as well as the Natal Botanical Institute have been working with healers since the early 1990s to set up more sustainable means of preserving indigenous flora. This has been pursued largely through the setting up of nurseries and cooperative *muthi* (medicine) gardens. For more information see the epilogue.

38. Peter Eagles, head of Medicines Control Council, interview, July 15, 2002, Cape Town. Also interviews with healers from Western Cape Traditional Healing Association: Phillip Kubukeli, July 19, 2002, and Gladis Williams, August 2, 2002, both Cape Town.

39. Cultural brokers can either be individuals or groups. A growing body of literature has focused on the importance of cultural brokers in the invention, creation, and imagination of tradition, tribalism, and ethnicity in Africa. E. Hobsbawm and

T. Ranger, eds., *The Invention of Tradition* (Cambridge: Cambridge University Press, 1983); B. Anderson, *Imagined Communities: Reflections on the Origin and Spread of Nationalism* (London: Verso, 1983); Mudimbe, *Invention of Africa*; Vail, *Creation of Tribalism*; Ranger, "Invention Revisted."

40. For a good summary of what is some times referred to as the "constuctivist approach" see S. Cornell and D. Hartmann, *Ethnicity and Race: Making Identities in a Changing World* (Thousand Oaks, CA: Pine Forge Press, 1998); R. Jenkins, *Rethinking Ethnicity* (Thousand Oaks, CA: Sage, 1997).

41. See L. Vail, *The Creation of Tribalism in Southern Africa*; R. Morrell, ed., *Political Economy and Identities in KwaZulu-Natal* (Durban: Indicator Press, 1996); and B. Carton, J. Laband, and J. Sithole, eds., *Zulu Identities: Being Zulu, Past and Present* (Pietermaritzburg University of KwaZulu-Natal Press, 2008).

42. For a discussion on the heterogeneity of this group see R. Ebr.-Vally, *Kala Pani* (Cape Town: Kwela, 2001). "Zanzibaris" were descendants of Muslim Africans from Northern Mozambique and Zanzibar whose ancestors had been seized by Arab and French slavers, captured by the British antislaving patrol and "released" to indenture in the Durban area. Because they were Muslim, they were treated as a distinct African community and later categorized as "other Asians" during the apartheid era. While Zanzibaris live/d in Indian townships and went to Indian schools, they have retained an identity distinct from many of their Indian neighbors, and intermarriage has been limited. For an excellent overview on this community, see Z. Seedat, "The Zanzibaris in Durban: A Social Anthropological Study of the Muslim Descendants of African Freed Slaves Living in the Indian Area of Chatsworth" (master's thesis in African Studies, University of Natal, Durban), 1973.

43. This may not seem like a particularly new or innovative notion, and in fact anthropologists have been exploring and challenging the notion of culture as a bounded entity since the early 1980s. In *Tradition*, Shils talks about traditions as unvariagated raindrops on a window pane that run together and part into new streams of water. Most examinations of cultural encounters, however, have focused on the encounters of dominant and dominated peoples and the process of globalization rather than the encounters of subaltern groups themselves. For example: H. Bhabha, *The Location of Culture* (London: Routledge, 1994); G. Baumann, *The Multicultural Riddle: Rethinking National, Ethnic, and Religious Identities* (NewYork: Routledge, 1999); T. Burke, *Lifebuoy Men, Lux Women* (Durham: Duke University Press, 1996); M. Gomez, *Exchanging Our Country Marks* (Chapel Hill: University of North Carolina Press, 1998); J. Hutnyk, *Critique of Exotica Music, Politics and the Culture Industry* (London: Pluto Press, 2000); D. Kertzer, *Ritual, Politics, and Power* (New Haven: Yale Unversity Press, 1988); A. McClintock, *Imperial Leather* (New York: Routledge, 1995); R. Takaki, *A Different Mirror* (Boston: Little, Brown and Co., 1993); J. Tomlinson, *Cultural Imperialism: A Critical Introduction* (Baltimore: Johns Hopkins University Press, 1991).

44. R. D. G. Kelley, "People in Me," *Colorlines* 1, no. 3 (1999): 5–7; "Author Criticizes Multiculturalism," *Georgetown Voice* (April 26, 2001); V. Prashad, "Afro-Dalits of the Earth, Unite!" *African Studies Review* 43, no. 1 (2000): 189–201. V. Prashad, *Everybody Was Kungfu Fighting* (Boston: Beacon Hill Press, 2001).

45. This was the case with more than eleven hundred people in 1985. Minister of Home Affairs, RSA, Mr. Stoffel Botha disclosed that these changes had occurred in 1985. KwaMuhle Museum Exhibit 1998.

46. Indians were generally categorized as "Asians," which meant they could not live or own property in areas categorized as "Colored." Specific examples can be found in U. Dhupelia-Mesthrie, *From Cane Fields to Freedom* (Cape Town: Kwela, 2001), 31.

47. An excellent history of apartheid's health system is L. Baldwin-Ragaven, *An Ambulance of the Wrong Color—Health Professionals, Human Rights and Ethics in South Africa* (Cape Town: University of Cape Town Press, 1999).

48. "Report of the National European-Bantu Conference, Cape Town, February 6–9, 1929"; and A. Matheson, "Social Medicine among the Kraals of Natal: Modern Methods of Health Propaganda Oust the Old-Time Witchdoctors of Polela (n. Bulwer)" (1947, pamphlet #27829). Both housed at Killie Campbell Archives, Durban, RSA.

49. This information was garnered by interviews with traditional healers and does not refer to parasitic worms that people may see in faeces. E. Green, *Indigenous Theories of Contagious Disease* (Walnut Creek, CA: AltaMira Press, 1999), 92–96.

50. In a multicultural and medically plural community where ailments cross boundaries and there are numerous medical therapies to choose from, how is it that patients and their families choose to attend one practitioner over another? Generally, the illness and circumstances surrounding its emergence determine its nature, and thus how and from whom to seek treatment. Anthropological works focusing on medical pluralism have provided useful models for thinking about the interaction of different healing systems and the therapeutic choices people make. Janzen is responsible for launching this investigation of a patient-centered approach to medical anthropology and understanding how it is that patients and their families make medical decisions in a multitherapeutic society. See J. Janzen, *The Quest for Therapy in Lower Zaire* (Berkeley and Los Angeles: University of California Press, 1978).

51. J. Parle and F. Scorgie, "Bewitching Zulu Women: Umhayiso, Gender and Witchcraft in Natal," unpublished paper presented at the African Studies Association, Houston, Texas, November 2001.

52. R. Simons and C. Hughs, eds., *The Culture-Bound Syndromes* (Boston: D. Reidel Publishing, 1985).

53. For a good discussion on the cultural impact and ethical considerations following the introduction of Prozac to the American market, particularly with regard to increased diagnoses of OCD and depression, see P. Kramer, *Listening to Prozac* (New York: Viking Press, 1993).

54. For early discussions on medical pluralism see J. Janzen, *Quest for Therapy*; J. Ford, *The Role of Trypanosomiasis in African Ecology: A Study of the Tsetse Fly Problem* (Oxford: Clarendon Press, 1971); and C. Leslie, "Pluralism and Integration in the Indian and Chinese Medical Systems," in *Culture, Disease, and Healing: Studies in Medical Anthropology*, ed. D. Landy (New York: Macmillan, 1977), 511–17.

55. Matheson, "Social Medicine," 8.

56. Seedat, "The Zanzibaris in Durban."

57. Green, *Indigenous Theories*, 168–69.

58. African nurse Katie Makanya acting as a mediator between biomedicine and local understandings of wellness convinced African women not to feed their newborns food by telling them that this would interfere with the effectiveness of the medicine. M. McCord, *The Calling of Katie Makanya* (Cape Town: David Philip, 1995), 177.

59. In addition to a more general call for testing the efficacy of traditional medicines, has been a call to test *ubhejane*, an herbal mixture sought out by many HIV/AIDS patients seeking treatment or an alternative to antiretrovirals. "South Africa: Small clinic at centre of debate over traditional medicine" in *IRIN, Humanitarian News and Analysis*. United Nations Office for the Coordination of Humanitarian Affairs, May 1, 2006. http://www.irinnews.info/S_report.asp?ReportID=53090&SelectRegion=Southern_Africa (last accessed March 6, 2007).

60. This is Green's primary argument in *Indigenous Theories*.

61. The historical exceptions include work by the following historians: Feierman, "Struggles for Control: The Social Roots of Health and Healing in Modern Africa," *African Studies Review* 28, no. 1 (1985): 73–147, and *Peasant Intellectuals*; M. Lyons, *The Colonial Disease: A Social History of Sleeping Sickness in Northern Zaire, 1900–1940* (Cambridge: Cambridge University Press, 1992); G. Waite, *A History of Traditional Medicine and Health Care in Precolonial East-Central Africa* (New York: Edwin Mellen Press, 1992); J. Janzen and E. Green, "Continuity, Change, and Challenge in African Medicine," in *Medicine across Cultures: History of Non-Western Medicine*, ed. D. Vasiljevic et al. (Boston: Kluwer Academic Publishers, 2003).

62. M. Hunter, *Reaction to Conquest* (London: Oxford University Press, 1936); E. Krige, *The Social System of the Zulus* (London, 1936); M. Gluckman, "Zulu Women in Hoecultural Ritual," *Bantu Studies* 9 (1935): 255–71; Evans-Pritchard, *Witchcraft, Oracles*.

63. *Africa, Journal of the International Institute of African Languages and Cultures* (1935) included the contributions of E. E. Evans-Pritchard, A. Richards, G. Orde Browne, C. Clifton Roberts, and F. Melland, who wrote on the impact of colonialism on African beliefs and witchcraft and the growing syncretism of African beliefs. A number of anthropologists began focusing on the impact of urbanization on African culture and life, e.g., L. Kuper, *African Bourgeois* (New Haven, CT: Yale University Press, 1965). In terms of its application

to African therapeutics, only one later study stands out: B. du Toit, "The Isangoma: An Adaptive Agent among Urban Zulu," *Anthropological Quarterly* 44, no. 2 (1971): 51–65.

64. R. Shaw, "The Politics and the Diviner: Divination and the Consumption of Power in Sierra Leone." *Journal of Religion in Africa* 26 (February 1996): 30–55; N. Gray, "Witches, Oracles, and Colonial Law: Evolving Anti-witchcraft Practices in Ghana, 1927–1932," *International Journal of African Historical Studies* 34, no. 2 (2001): 339–63; N. Gray, "Independent Spirits: The Politics of Policing Anti-witchcraft Movements in Colonial Ghana, 1908–1927," *Journal of Religion in Africa* 35, no. 2 (2005): 139–58; P. Geshiere, *The Modernity of Witchcraft* (Charlottesville: University Press of Virginia, 1997).

65. Green, *Indigenous Theories*, 12.

66. N. Oudshoorn, *Beyond the Natural Body: An Archeology of Sex Hormones* (London: Routledge, 1994), and E. Martin, "The Egg and the Sperm: How Science Has Constructed a Romance Based on Stereotypical Male-Female Roles," *Signs* 16, no. 3 (Spring 1991): 485–501. D. Wylie looks at the cultural construction of famine in *Starving on a Full Stomach* (Charlottesville: University of Virginia Press, 2001). Likewise, M. Vaughan examines the ways psychiatry contructed the pathological African mind in *Curing Their Ills: Colonial Power and African Illness* (Stanford: Stanford University Press, 1991).

67. Feierman, "Struggles," 110.

68. Ibid, 73–147.

69. Political economists of health include R. Packard, *White Plague, Black Labor* (Berkeley and Los Angeles, University of California Press, : 1989); T. Falola and D. Ityavyar, *The Political Economy of Health in Africa* (Athens: Ohio University Press, 1992); and many chapters in Feierman and Janzen, *Social Basis*.

70. Foucauldian scholars include R. Macleod and M. Lewis, eds., *Disease, Medicine, and Empire: Perspectives on Western Medicine and the Experience of European Expansion* (New York: Routledge, 1988); T. Ranger and P. Slack, eds., *Epidemics and Ideas: Essays on the Historical Perception of Pestilence* (New York: Cambridge University Press, 1992); Vaughan, *Curing Their Ills*; D. Arnold, ed., *Imperial Medicine and Indigenous Societies* (Manchester: Manchester University Press, 1988); M. Echenberg, *Black Death, White Medicine* (Portsmouth, NH: Heinemann, 2002); A. Butchard, *The Anatomy of Power: European Constructions of the African Body* (London: Zed Books, 1998); Wylie, *Starving*. For the African American context see S. Fett, *Working Cures* (Chapel Hill: University of North Carolina Press, 2002).

71. Those scholars influenced by Gramsci include: Marks, *Divided Sisterhood*; J. Iliffe, *East African Doctors: A History of the Modern Profession* (Cambridge, Cambridge University Press, 1998); N. Rose Hunt, *A Colonial Lexicon* (Durham: Duke University Press, 1999); H. Bell, *Frontiers of Medicine in the Anglo-Egyptian Sudan* (Oxford: Clarendon Press, 1999). Also see many of the chapters in Erst, *Plural Medicine*.

72. Dr. Adam Smith, who visited southeastern Africa in 1832, replicated parts of Fynn's and Farewell's journals. Furthermore the convention of the early to mid-nineteenth century seemed to permit the copying of large tracts of previously published material to bolster one's own travel memoirs, e.g., Rev. Shooters (1857) quotes both Fynn and Isaacs at length.

73. Nathaniel Isaacs to Henry Fynn, December 10, 1832, published in *Africana Notes and News* 18 (1968–69): 67. As quoted in B. Worger, "Clothing Dry Bones: The Myth of Shaka," *Journal of African Studies* 6, no. 3 (1979): 145.

74. Hamilton, *Terrific Majesty*, 144.

75. *Commission to inquire into the past and present state of the Kafirs. Proceedings and report of the Commission appointed to inquire into the past and present state of the Kafirs in the district of Natal, and to report upon their future government and to suggest such arrangements as will tend to secure the peace and welfare of the district: for the information of His Honour Lieutenant-Governor Pine: 1852–3* (Pietermaritzburg, 1879) (hereafter, *Natal Native Commission*), 1852 testimony of Fynn, 55–60.

76. O. Olumwullah, *Dis-ease in the Colonial State: Medicine, Society and Social Change among the AbaNyole or Western Kenya* (Westport, CT: Greenwood Press, 2002), see chapter 1; and A. Butchart, *The Anatomy of Power*.

77. J. McCord, "Medical Missionary Work in Africa," *Christianity and the Native of South Africa: A Yearbook of South African Missions* (1928); J. McCord, "Some Sidelights on the House that Jim Built," Killie Campbell, Pamphlet 362; MACC, 1929; Rev. Tyler, *Forty Years Among the Zulus* (Boston, 1891). As compared to: J. McCord, "A Zulu Witch Doctor and Medicine Man," *South African Medical Record* 16, no. 8 (1918): 116–22; Rev. Callaway, *The Religious System of the Amazulu in Their Own Words* (London, 1870); F. Hale, ed., *Norwegian Missionaries in Natal and Zululand: Selected Correspondence, 1844–1900* (Cape Town: Van Riebeeck Society, 1997).

78. Fynn tells the 1852 Natal Native Commission that he had studied the craft of healers for twenty-eight years. His own detailed descriptions in his Diary, the Commission, and knowledge of African medicine as attested by William Basely in the James Stuart archives all seem to attest to this expertise.

79. Bryant wrote several pieces on African healers, including "The Zulu Cult of the Dead," *Man* 17, no. 95 (1917): 140–45, and *Zulu Medicine and Medicine Men* (1909; repr., Cape Town: C. Struit, 1966), both of which are scholarly pieces, as opposed to his later works, which are written for a popular audience and in which he readily admits to fluffing out details for the sake of entertainment.

80. H. Callaway's *The Religious System of the Amazulu in Their Own Words* (London, 1870) (1948) attributes over half of its testimony to Mpengula Mbande, a member of Callaway's church, but also includes the evidence of Mbande's brother, who was trained as a healer, and individuals fleeing the *mfecane* from the Transkei and northern Zululand. N. Etherington, "Missionary Doctors and African Healers in Mid-Victorian South Africa," *South African Historical Journal* 19 (1987): 81–82.

81. For an analysis of sources on this topic see K. Flint, "Perceptions of African Healers in the Nineteenth and Twentieth Centuries," in "Negotiating a Hybrid Medical Culture: African Healers in Southeastern Africa from the 1820s to 1940s" (PhD diss., University of California, Los Angeles, 2001), 10–47.

82. The accused for instance could also ask for another healer to try their case, or the king might intervene on their behalf. For examples one can look at some of the testimony in the James Stuart Archives. C. B. Webb and J. B. Wright, eds., *The James Stuart Archive of Recorded Oral Evidence Relating to the History of the Zulu and Neighbouring Peoples*, 5 vols. (Pietermaritzburg: University of Natal Press, 1976, 1979, 1982, 1986, 2001) (hereafter referred to as JSA). JSA, 2:214; JSA, 3:308. Also see R. H. Reyher, *Zulu Woman: The Life Story of Christina Sibiya* (1948; repr., New York: Feminist Press at CUNY, 1999), 149.

83. Although it was originally scripted in long- and shorthand with sections in Zulu and English, most authors, including myself, have preferred to use the translated, edited, and published version of the archives largely prepared by J. Wright. For a critique of the Stuart Archives, see J. Cobbing, "A Tainted Well: The Objectives, Historical Fantasies and Working Methods of James Stuart, with Counter-argument," *Journal of Natal and Zulu History* 11 (1988): 115–54. Also see counter-critique by J. Wright in "Making the James Stuart Archive," *History in Africa* 23 (1996): 333–50, and an analysis of this collection by C. Hamilton in *Terrific Majesty*.

CHAPTER 1: HEALING THE BODY

1. J. Stuart, ed., *The Diary of Henry Francis Fynn* (Pietermaritzburg: Shuter and Shooter, 1950), 42–43.

2. His diary includes additional notes written on this topic, and he tells the 1852 Natal Native Commission that he has studied healers for the past twenty-eight years. *1852 Natal Native Commission*, evidence of Fynn, 56. For an example of Fynn healing Africans, see Stuart, *Diary of Fynn*, 66, 77, 84–88. Healing of Mr. Bazley: JSA, 1:60.

3. A. T. Bryant, *The Zulu People as They Were before the White Man Came* (Pietermaritzburg: Shuter and Shooter, 1949), 377.

4. Fynn mentions purchasing a remedy in Natal Native Commission, 1852, 52, as do J. Medley Wood in "Native Herbs: Medicinal and Otherwise," 1898 *Natal Almanac*, 264, and Callaway, *Religious System*, 417–18.

5. Referring to current practices, Queen Ntuli told the *Wall Street Journal* that healers have their own means of dealing with intellectual property rights, thus if a healer wanted to purchase a remedy from another healer, she would pay a cow and then perform rituals to ensure that their ancestors agreed to the transfer. M. Phillips, "Persuading Africans to Take Their Herbs with Some Antivirals," *Wall Street Journal*, May 5, 2006.

6. J. Janzen, *Ngoma: Discourses in Healing* (Berkeley: University of California Press, 1992). Janzen's work is based partly on historical linguistics and he uses M.

Guthrie's list of common word cognates and roots found in M. Guthrie, *Comparative Bantu: An Introduction to the Comparative Linguistics and Prehistory of the Bantu Languages*, 4 vols. (Farnborough: Gregg, 1967–71). More recently see Janzen's responding chapter in *The Quest for Fruition through Ngoma: The Political Aspects of Healing in Southern Africa*, ed. R. Van Dijk, R. Reis, and M. Spieranburg (Athens: Ohio University Press, 2000).

7. For a history of the interaction of these groups see M. Hall, *Farmers, Kings, and Traders: The People of Southern Africa 200–1860* (Chicago: University of Chicago Press, 1990).

8. For an example of historical linguistics applied to African medicine, see Waite, *History*.

9. *JSA*, 1:42.

10. Though iron smithing was common throughout southeastern Africa, the Cube were recognized as excelling in this skill. *JSA*, 1:42; *JSA*, 4:14.

11. B. Van Wyk, B. Van Oudtshoorn, and N. Gericke, *Medicinal Plants of South Africa* (Pretoria: Briza, 1997), 102.

12. This may in fact be due to overharvesting, but clearly the ecological diversity meant that certain plants could only be grown in certain regions and thus different groups would be prone to specialize. Ibid., 272.

13. *JSA*, 3:263; *JSA*, 4:37.

14. For Zulu: *JSA*, 3:25; *JSA*, 2:84; for Mpondo, *JSA*, 5:308; and for tobacco, *JSA*, 2:24.

15. For the Zolo, *JSA*, 3:134–37, and *JSA*, 2:84; for the Tshangala, *JSA*, 2:84; for the Swazi, *JSA*, 1:67.

16. Bryant, *Zulu People*, 376, 244.

17. C. Hamilton, ed., *The Mfecane Aftermath* (Johannesburg: Witwatersrand University Press, 1995).

18. Thongoland: J. Guy, *Destruction of the Zulu Kingdom: Civil War in Zululand, 1879–1884* (London: Longman, 1979), 17. Mpondo: *JSA*, 5:308.

19. For detailed ecology of this area see J. Guy, "Ecological Factors in the Rise of Shaka and the Zulu Kingdom," in *Economy and Society in Pre-industrial South Africa*, ed. S. Marks and A. Atmore (London: Longman, 1980), 102–19.

20. Wylie, *Starving*, 42–52; for quotation see page 47.

21. For information on food crops see Stuart, *Diary of Fynn*, 304–6; N. Isaacs, *Travels and Adventures in Eastern Africa* (Cape Town: C. Struik, 1970), 306. Letter from Scheuder dated 1846. F. Hale, *Norwegian Missionaries in Natal and Zululand: Selected Correspondence, 1844–1900* (Cape Town: Van Riebeeck Society, 1997), 29; *JSA*, 5:174; *JSA*, 5:377. See J. Chase, *The Natal Papers: A Reprint of All Notices and Public Documents Connected with That Territory Including a Description of the Country and a History of Events from 1498 to 1843* (Cape Town: C. Struik, 1968), 9, 32. For nutritional information and discussion about indigenous plants see B. van Wyk and N. Gericke, *People's Plants* (Pretoria: Briza, 2000), 16, 28, 114.

22. F. W. Fox and W. Stone, "The Anti-scorbutic Value of Kaffir Beer," *South African Journal of Medical Sciences* 1 (1938): 7–14.

23. Guy, *Destruction*, 7.

24. Isaacs, *Travels and Adventures*, 306.

25. O. H. Spohr, *The Natal Diaries of Dr. W. H. I. Bleek, 1855–1856* (Cape Town: Balkema, 1965), 31, 35.

26. The 1904 Land Delimitation Commission opened Zululand to white settlement.

27. B. van Wyk and N. Gericke, *People's Plants* (Amadumbe, 82, maize, 16). Also for maize, P. Kirby, ed., *Andrew Smith and Natal: Documents Relating to the Early History of That Province* (Cape Town: Van Riebeeck Society, 1955), 169; Chase, *Natal Papers*, 9.

28. Stuart, *Diary of Fynn*, 45.

29. From R. Plant, "Notice of a Excursion in Zulu Country," in D. Mc-Cracken and P. McCracken, *The Natal: The Garden Colony* (Sandton: Frandsen, 1990), 7. A. T. Bryant also claimed that malaria occurred mainly within eight miles of the coast (*Zulu People*, 121). Typical of the colonial encounter, transportation corridors contributed greatly to the rapid spread of disease and ecological change. Malaria spread farther inland in Natal after the railways were built (beginning in 1876). J. McCord claims that malaria spread even further into Natal with the 1904 railroad extension to Stanger into malarial areas of Zululand as anopheles mosquitoes were locked in train coaches. Furthermore, railways not only transported malarial persons and mosquitoes inland, but their embankments created new breeding areas for mosquitoes. J. McCord, *My Patients Were Zulus* (New York: Rinehart, 1951), 249; Marks, *Divided Sisterhood*, 40.

30. Hale, *Norwegian Missionaries*, 35.

31. Bryant, *Zulu People*, 121, and Stuart, *Diary of Fynn*, 281.

32. For gonorrhea see K. Kiple, ed., *Cambridge World History of Human Disease* (New York: Cambridge University Press, 1993), 449. For a discussion of indigenous diseases of Africa and those which were introduced with colonialism, see "Diseases of Sub-Saharan Africa to 1860" and "Disease Ecologies of Sub-Saharan Africa," both in Kiple. This was due in part to grazing cattle at higher elevations where tsetse fly did not exist.

33. Guy, *Destruction*, 14.

34. G. Hartwig and D. Patterson, *Disease in African History* (Durham, NC: Duke University Press, 1978), 3–21.

35. E. H. Cluver, *Public Health in South Africa* (Johannesburg: Central News Agency, 1959); Bryant, *Zulu Medicine*, 24; J. Brain, "Health and Disease in White Settlers in Colonial Natal," *Natalia* 15 (1985). Also see *Cambridge World History of Human Disease*.

36. Hale, *Norwegian Missionaries*, 29.

37. Kirby, *Andrew Smith*, 49.

38. J. H. S. Gear, "The History of Virology in South Africa," *South African Medical Journal* (October 11, 1986), supplement, 7–10. The indigenousness of

smallpox is still debated; see M. Dawson, "Socioeconomic Change and Disease: Smallpox in Colonial Kenya, 1880–1920," in *Social Basis*, ed. Feierman and Janzen, 90–103. Also see *Cambridge World History of Human Disease*, 296.

39. Bryant, *Zulu People*, 122. *JSA*, 2:143, evidence of Mahungane.

40. P. Harries, *Work, Culture, and Identity* (London: James Currey, 1994), 12.

41. J. Y. Gibson, *The Story of the Zulus* (Pietermaritzburg: Davis and Sons, 1903), 108.

42. Guy, *Destruction*, 7, 32.

43. K. Kiple, *Cambridge World History of Human Disease*, 295.

44. Spohr, *Natal Diaries*, 17, 31, 34, 53. Oftebro, a medical missionary, claimed that tuberculosis in cattle was a severe problem in the kingdom by 1851. It is unclear whether he was referring to the same disease that Bleek referred to four years later. Hale, *Norwegian Missionaries*, 49.

45. Guy, *Destruction*, 15.

46. Hale, *Norwegian Missionaries*, 201.

47. G. Waite, "Public Health in Precolonial East-Central Africa," in Feierman and Janzen, *Social Basis*, 213.

48. While the authors are referring specifically to ngoma healing institutions, which were and are practiced in the area of KwaZulu-Natal, such definitions aptly apply to the wide variety of healers particularly in the Zulu kingdom. Van Dijk et al., *Quest*, 6.

49. Ibid., 166.

50. For other examples in Africa see Waite, *History*; Feierman, "Struggles"; D. Maier, "Nineteenth-Century Asante Medical Practices," *Comparative Studies in Society and History* 21 (1979): 63–81.

51. *JSA*, 3:185.

52. Evidence of Robert Plant, from McCracken and McCracken, *Natal*, 7. For other environmental notes see J. Pieres, "Introductory Notes," in *Before and After Shaka: Papers in Nguni History* (Grahamstown: Institute of Social and Economic Research: Rhodes University, 1981), 2–5. Tsetse fly: Spohr, *Natal Diaries*, 59.

53. Spohr, *Natal Diaries*, 15–16.

54. This is a change from earlier times when larger groups of people resided in one kraal for protection. Rev. L. Grout, *Zululand* (1864), 96–97; *JSA*, 1:201.

55. *JSA*, 5:180.

56. Stuart, *Diary of Fynn*, 291.

57. *JSA*, 1:176; *JSA*, 4:94.

58. *JSA*, 5:183.

59. Cetshwayo's herds: C. Vijn, *Cetshwayo's Dutchman* (New York: Negro Universities Press, 1968), 57.

60. Stuart, *Diary of Fynn*, 277.

61. *JSA*, 5:164.

62. D. Leslie, *Among the Zulus and Amatongas* (1875; repr., New York: Negro Universities Press, 1969), 183; and Spohr, *Natal Diaries*, 70.

63. *JSA*, 5:342.

64. C. Barter, *Alone among the Zulus: The Narrative of a Journey through the Zulu Country, South Africa* (1866; repr., Pietermaritzburg: University of Natal Press, 1995), 100.

65. *JSA*, 4:41; *JSA*, 2:56.

66. *JSA*, 3:141. *Zulu Woman*, 126.

67. *JSA*, 4:374; quote is from Kirby, *Andrew Smith and Natal*, 74.

68. The rhetoric of racial hygiene develops later in the nineteenth century; see M. Swanson, "The Sanitation Syndrome: Bubonic Plague and Urban Native Policy in the Cape Colony, 1900–1909," *Journal of African History* 18, no. 3 (1979): 387–410; J. and J. Comaroff, "Medicine, Colonialism, and the Black Body," 224–26. For cultural construction of hygiene see T. Burke, *Lifebuoy Men, Lux Women* (Durham: Duke University Press, 1996), chapter 1.

69. *JSA*, 4:375.

70. Stuart, *Diary of Fynn*, 292–94.

71. J. A. Mitchell, "Health, Growth and Welfare of the Native Peoples," in *Report of the National European-Bantu Conference* (Cape Town, 1929).

72. Stuart, *Diary of Fynn*, 292–94. A. Butchard describes early Europeans' fascination with the Khoisan.

73. Stuart, *Diary of Fynn*, 70; G. Cory, *The Diary of Rev. Francis Owen* (Cape Town: Van Riebeeck Society, 1965), 56; T. B. Jenkinson, *Amazulu: The Zulus, Their Past History, Manners, Customs and Language: Natal 1873–1879* (London, 1882), 54. *Report of the Commission of Leprosy*, 3.

74. *JSA*, 1:45; *JSA*, 3:162.

75. Maier, "Nineteenth-Century Asante Medical Practices," 68–69. Dr. Gumede, then Secretary for Health in northern Zululand, refers to the continued belief in witchcraft as late as 1981–82 and its effect on local health. During this period, cholera prompted the Department of Health to require an outhouse for every homestead. Resistance to this requirement reflected not only the fear of umthakathis, who with access to a household outhouse may eliminate an entire family, but also reflected local customs that prevented fathers-in-law from sharing the same seat as their daughters-in-law or the wives of their sons. Consequently, many of the toilets built remained unused. M. V. Gumede, *Traditional Healers* (Cape Town: Skotaville, 1990), 52–54.

76. *JSA*, 3:175, 228; *JSA*, 4:43–44; Bryant, *Zulu People*, 662–64.

77. *JSA*, 4:307–10; Bryant, *Zulu People*, 664–68; M. Gluckman, "Zulu Women in Hoe Cultural Ritual," *Bantu Studies* 9 (1935): 255–71.

78. See http://www.library.unp.ac.za/paton/Nomkhubulwane.htm (accessed December 15, 2005). Patience Gugu Ngobese began this custom while an honors student at UKZN in 1996.

79. For two excellent articles on this topic, see S. Leclerc-Madlala, "Virginity Testing: Managing Sexuality in a Maturing HIV/AIDS Epidemic," *Medical Anthropology Quarterly* 15, no. 4 (2001): 533–52; and Scorgie, "Virginity Testing," 55–75.

80. Stuart, *Diary of Fynn*, 286; Spohr, *Natal Diaries*, 15, 54, 60.

81. Hale. *Norwegian Missionaries*, 37. Also see Wylie, *Starving*, 39–55.

82. Hale, *Norwegian Missionaries* (from Scheuder in 1846 and Sanne in 1897), 29, 200; in a later article, mention is made of various herbs that Africans eat with anthrax-infected meat. Dr. J. Lewsi, Assistant Government Analyst, and O. F. Gibbs, Pathological Assistance, Government Laboratory, Bloemfontein, "Native Preventatives of Anthrax Infection," *South African Medical Record* 19, no. 12 (June 25, 1921).

83. *JSA*, 4:316.

84. *JSA*, 4:363; *JSA*, 1:233.

85. S. J. Maphalala, *Aspects of Zulu Rural Life during the Nineteenth Century* (Empangeni: University of Zululand, 1985), 5.

86. *JSA*, 3:155.

87. *JSA*, 4:363.

88. *JSA*, 2:89. Fynn called upon by Tshaka to treat people within his kraal and reprimanded for treating commoner. *Dairy of Fynn*, 77, 277, 132; likewise Scheuder and Gundersen called by Mpande and Cetswayo to attend to sick people within the king's kraal as well as important men of the kingdom. Hale, *Norwegian Missionaries*, 21, 35–37, 70–71, 75.

89. G. Cory, *Diary of Owen*, 83, 87.

90. Isaacs, *Travels and Adventures in Eastern Africa* (Cape Town: C. Struik, 1970), 48–49; *JSA*, 1:278; the cause of the banning of the metal differs in this account. *JSA*, 1:23–24.

91. A. Booth, ed., *Journal of Rev. George Champion, American Missionary in Zululand, 1835–9* (Cape Town: C. Struik, 1967), 25.

92. *JSA*, 4:70; *JSA*, 1:69–72. V. Sibeko, "Imilingo nemithi yokwelapha phakathi kwabantu" (Magic and healing medicines of the people), *Natal Native Teacher's Journal* 11, no. 3 (1932).

93. *JSA*, 4:120, 271; *JSA*, 1:325–28; Bryant, *Zulu People*, 511–21.

94. Grout, *Zululand*, 156–57; Callaway, *Religious System*, 280n39.

95. This in mentioned from Fynn's Diary up to Callaway's work in the 1870s. Stuart, *Diary of Fynn*, 278, 280.

96. J. L. Dohne, *A Zulu-Kafir Dictionary Etymologically Explained, with Copious Illustrations and Examples, Preceded by an Introduction on the Zulu-Kafir Language* (Cape Town: Farnborough, 1857).

97. Inkatha: Bryant, *Zulu People*, 469–71, 476, 517.

98. Berglund claims that isanusi were men and diviners of the king. A. Berglund, *Zulu Thought Patterns and Symbolism* (New York: Holmes and Meier, 1976), 185. While this may be how Berglund's respondents refer to the isanusi, primary sources such as the James Stuart Archives and early European writers use these terms interchangeably, for men and women, and for important and less important healers. It seems the main difference in these terms is regional, something that current day healers also confirmed. M. Mthemjwa, interview, October 16, 1998, Durban. J. Mbuyazi, interview, November 29, 1998, Mthubathuba.

99. Callaway, *Religious System*, 299–304.

100. Quote from Callaway, *Religious System*, 321–22; also this type of healer is mentioned in *JSA*, 4:321.

101. *JSA*, 1:342; *JSA*, 4:317; *JSA*, 3:177–78. Evidence given by Dlhozi, Mkanda, and Ndukwana. Dlhozi's father had consulted an umlozikazana; Samuelson, *Long, Long Ago* (Durban: Knox, 1929), 393.

102. Callaway, *Religious System*, 348–74; *JSA*, 1:342.

103. Bula-ing: *JSA*, 1:324.

104. This is described in great detail in chapter 3.

105. Grout, *Zululand*, 145; *JSA*, 4:223.

106. *Natal Native Commission 1552*, vol. 5, 55, evidence of Fynn.

107. Stuart, *Diary of Fynn*, 280; S. G. Campbell, "Zulu Witchdoctors and Experiences amongst the Zulus," *Glasgow Medical Journal* (October 1888); Bryant, *Zulu People*, 377.

108. Stuart, *Diary of Fynn*, 280.

109. Samuelson, *Long, Long Ago*, 293.

110. *JSA*, 2, Gunderson's letter from 1876, 206.

111. Hale, *Norwegian Missionaries*, 71; *JSA*, 3:169.

112. Spohr, *Natal Diaries*, 38. Bleek noted in 1855 that Africans did not like dissection and that he doubted he could secure a corpse to dissect.

113. In Zulu nomenclature only a handful of words describe internal organs. There are more than 250 words in Zulu for human body parts, but the earliest rendering that we have comes from Killie Campbell Collection, Guy Essery #3, "Zulu Medicine" (1927); M. V. Gumede, *Traditional Healers*, 109–13.

114. Bleek tells us that an inyanga told him about dissection. Spohr, *Natal Diaries*, 33.

115. *JSA*, 1:10; *JSA*, 1:53; *JSA*, 1:7, 195; *JSA*, 5:40, 90. Worger offers an alternative explanation to Tshaka's killing of the pregnant woman. Worger, "Clothing Dry Bones."

116. Dr. Campbell speaks of the hostility he faced. Africans told him, "Is not the man dead already, why do you want to kill him again?" S. G. Campbell, "Zulu Witchdoctors."

117. *JSA*, 3:174.

118. Dohne, *Zulu-Kafir Dictionary*. Other descriptions include Bryant's term for all fevers and natural diseases (*Zulu Medicine*, 51–52); An informant in *JSA* says that it is "a fever, not serious disease." *JSA*, 3:175, 184, (1902); C. Doke et al., *Zulu-English Dictionary* (1948; repr., Johannesburg: Witwatersrand University Press, 1990), 409; Gumede says that this category may include measles, smallpox, mumps, or whooping cough, or where a speedy recovery is expected (*Traditional Healers*, 45).

119. Booth, *Journal of Rev. George Champion*, 80; A. T. Bryant, A *Zulu-English Dictionary* (Pinetown: RSA, 1905), 347; *JSA*, 3:174.

120. S. G. Campbell, "Zulu Witchdoctors." Here he is talking of African healers.

121. Bryant, *Zulu Medicine*, 52; H. Ngubane, *Body and Mind in Zulu Medicine: An Ethnography of Health and Disease in Nyuswa-Zulu Thought and Practice* (London: Academic Press, 1977), 23.

122. Bryant, *Dictionary*, 290; *JSA*, 3:174; S. M. Molema, *The Bantu—Past and Present* (1920), 166; S. G. Campbell, "Zulu Witchdoctors."

123. Ngubane, *Body and Mind*, 23. Both Bryant and Colenso record in 1905 that the isiZulu word *ufuzo* stands for resemblance. Bryant, *Dictionary*; Rev. Colenso, *Zulu-English Dictionary* (Natal: Vause, Slatter, 1905).

124. Observing the effects of old categorizations on current medical practices. The medicines used to cure diseases of the umkhuhlane class are believed to be potent and effective in themselves. They are therefore not ritualized. Ngubane, *Body and Mind*, 23.

125. J. Stuart, "Zulu Beliefs and Superstitions," Stuart Paper File 2, KCM 1046, item 16. Killie Campbell Library, hereafter KCL.

126. *JSA*, 3:174; *JSA*, 4:302.

127. Isaacs, *Travels and Adventures*, 232.

128. *JSA*, 3:174.

129. *JSA*, 4:75.

130. Failure to engage in this ritual resulted in *iliqungo*, an insanity particular to this condition. R. Samuelson, *King Cetewayo Zulu Dictionary* (Durban: Commercial Printing, 1923), 556; Bryant, *Dictionary*, 549; *JSA*, 4:217; *JSA*, 5:166, 171.

131. *JSA*, 4:325.

132. *Natal Native Commission 1852*, V, 64, evidence of Fynn.

133. *JSA*, 1:228.

134. M. Hunter, *Reaction to Conquest: Effects of Contact with Europeans on the Pondo of South Africa* (London: Oxford University Press, 1936).

135. *JSA*, 4:302, 325.

136. *JSA*, 3:174.

137. *JSA*, 3:172; also see Callaway, *Religious System*, 371.

138. *JSA*, 2:174.

139. *JSA*, 3:181.

140. *JSA*, 3:182.

141. It is unclear whether these are Owen's assumptions about gender. J. Bird, *The Annals of Natal, 1495 to 1845* (Cape Town: C. Struik, 1965), 1:332.

142. *The Collector*, no. 209, as quoted in Krige, *Social System of the Zulus* (Pietermaritzburg: 1950), 321.

143. Bryant, *Dictionary*; Callaway, *Religious System*, 351n6.

144. *JSA*, 5:183.

145. *JSA*, 1:210.

146. *JSA*, 4:281.

147. *Natal Native Commission 1852*, V, 56, evidence of Fynn.

148. Stuart, *Diary of Fynn*, 232.

149. This term appears in Dohne, *A Zulu-Kafir Dictionary*, as *umbetelelo*; the more common word used for love charms today is *ubulawu*. Bryant, *Dictionary*.

150. Dr. McCord relates the physical affects of a love charm administered on both daughter and nurse. In this case both fell ill with a high fever. J. McCord, "Native Witch Doctors and Healers," *South African Medical Record* 24 (1926).

151. Bryant, *Dictionary*, 33–34. Bryant's terms are slightly different, reflecting perhaps a regional variation of the name or perhaps Bryant's ear for Zulu words. The originals appearing in this dictionary, which vary, are *umanaye* and *unginakile*. Love charms, of course varied in their concoctions, depending on access to specific muthis.

152. Reyher, *Zulu Woman*, 55.

153. *Natal Native Commission* 1852, vol. 5, 68, evidence of Fynn.

154. B. Carton, *Blood from Your Children* (2000); M. Mahoney and J. Parle, "An Ambiguous Sexual Revolution? Sexual Change and Intra-generational Conflict in Colonial Natal," *South African Historical Journal* no. 50 (2004): 134–51.

155. Stuart, *Diary of Fynn*, 307; *Natal Native Commission*, 1852. Part 5:36, evidence of George Robinson from Natal Witness; Krige, *Social System of the Zulus*, 330; Callaway, *Religious System*, 287.

156. S. G. Campbell, "Zulu Witchdoctors," 2, 14; Bryant, *Dictionary*.

157. Medley Wood, "Native Herbs." Compilations of Xhosa herbs were collected earlier, beginning in the 1850s. McCracken and McCracken, *Natal*, 32.

158. Bryant, *Zulu Medicine*. The contents were originally published in the *Annals of the Natal Museum* (Pietermaritzburg: 1909). Almost ninety years later, botanist Anne Hutchings catalogued 1,032 medicinal plants, at least a third of which represent knowledge acquired through oral sources. This increase *probably* reflects both Hutchings's extensive research and the exchange and increase in plant and medicinal knowledge prompted by urbanization and organization of healers during the twentieth century. A. Hutchings, *Zulu Medicinal Plants: An Inventory* (Pietermaritzburg: University of Natal Press, 1996).

159. Bryant, *Zulu Medicine*, 57–58. Jovela: S. G. Campbell, "Zulu Witchdoctors," 14; umsizi: Callaway, *Religious System*, 287.

160. J. McCord, "Zulu Witch Doctor."

161. Ibid.

162. When Cetshwayo was stabbed, his izinceku or servants found a reed and blew water into the wound to wash it; they then covered it with bark from a mimosa (umuga tree). *JSA*, 4, no. 66, evidence of Ndabazezwe; Jenkinson, *Amazulu*, 66. Dr. Campbell also writes of wounds washed in water and covered in aloe (S. G. Campbell, "Zulu Witchdoctors," 11).

163. McCord, "Zulu Witch Doctor," 118.

164. Stuart, *Diary of Fynn*, 280.

165. Grout, *Zululand*, 157; Stuart, *Diary of Fynn*, 280; Bryant, *Zulu Medicine*, 31; S. G. Campbell, "Zulu Witchdoctors," 10.

166. Bryant, *Zulu Medicine*, 33.

167. Bryant, *Dictionary*, 517; Bryant, *Zulu Medicine*, 22.

168. Quote from Stuart, *Diary of Fynn*, 281; bleeding also written about by Grout, *Zululand*, 157; Booth, *Journal of Rev. George Champion*, 87; W. Holden, *The Past and Future of the Kaffir Races* (Cape Town: Struiker, 1963), 373.

169. *JSA*, 4:342, evidence of Ndukwana.

170. Stuart, *Diary of Fynn*, 66n2. Callaway, *Religious System*, 290; Holden, *Past and Future*, 377; S. G. Campell, "Zulu Witchdoctors," 12; and numerous references in *JSA*.

171. Based on review of yearly Natal Blue Books between 1870s and 1910 which commented on both Africa and European compliance with vaccination. Unlike Africans, Europeans could refuse vaccination based on a conscientious clause. Africans however had other means of showing disfavor with vaccination requests.

172. L. Govender, interview, December 17, 1998, Durban.

173. A. Gardiner, *Narrative of a Journey to The Zoolu Country, in South Africa* (London, 1836), 308.

174. M. Hewat, *Bantu Folklore: Medical and General* (Westport, CT: Negro University Press, 1979; first impression, 1906), 60. Killie Campbell, Guy Essary, #3, "Zulu Medicine" (1927) also mentions a similar practice for a dislocated ankle.

175. Holden, *Past and Future*, 377.

176. Stuart, *Diary of Fynn*, 281.

177. Ibid., 84.

178. Bryant, *Zulu Medicine*, 36.

179. Hutchings, *Zulu Medicinal Plants*, 259.

180. Stuart, *Diary of Fynn*, 281; *JSA*, 5:171.

181. Stuart, *Diary of Fynn*, 301.

182. Ibid., 307; J. McCord, "Zulu Witchdoctor."

183. Stuart, *Diary of Fynn*, 311; "International Exhibit from Natal," *Government Gazette* 14, no. 689 (January 14, 1862); isihlungu used instead, *JSA*, 4:95.

184. Isaacs mentions being treated similarly: Isaacs, *Travels and Adventures*, 64–65. Noted by unnamed author, *South African Medical Journal*, March 8, 1889, 167.

185. Bryant, *Zulu Medicine*, 38–39.

186. Ibid., 58.

187. Stuart, *Diary of Fynn*, 281.

188. Hale, *Norwegian Missionaries*, 71.

189. Leslie, *Among the Zulus*, 102; this practice was witnessed by both J.M. McCord and his nurse Katie Makanya, who each wrote about it. M. McCord, *Calling*, 176; J.M. McCord, "Some Sidelights on the House That Jim Built," Killie Campbell, Pamphlet 362, MACC, 1929, 20.

190. *JSA*, 4:342.

191. H. C. Lugg, *A Natal Family Looks Back* (Durban: T. W. Greggs, 1970), 117.

192. The earliest reference to African suturing I have found comes from 1896: H. E. Fernandez (Port Shepston, Natal), "The Extraordinary Vitality of the Native under Serious Injuries—(illustrated by a recent case)," *South African Medical Journal* 4, no. 2 (June 1896).

193. African medical historian Marinez Lyons, commenting on African therapeutics, states, "Such knowledge will allow us to discern more clearly the under-

lying explanations for what might otherwise appear to be purely irrational behavior of Africans." M. Lyons, *The Colonial Disease: A Social History of Sleeping Sickness in Northern Zaire, 1900–1940* (Cambridge: Cambridge University Press, 1992), 15.

194. K. Flint, "Negotiating a Hybrid Medical Culture," Appendix III, "Recorded Cases of Witchcraft and Administering Poison in Natal, 1883–1949," 235.

195. Stuart, *Diary of Fynn*, 84–88, 277.

196. Ibid., 278.

197. Ibid., 280.

CHAPTER 2: HEALING THE BODY POLITIC

1. *JSA*, 1:181–82, evidence of Jantshi. A similar story is supported by *JSA*, 4, evidence of Mgundeni ka Matshekana, 122–23. Kunene and Fuze say that Tshaka obtained muthi from a diviner that was meant only for a king. When Senzangakona saw Tshaka he felt overwhelmed being ill, which necessitated his return to the Zulu chiefdom, after which he died. M. Kunene, *Emperor Shaka the Great: A Zulu Epic* (London: Heinemann, 1979), 68–70. M. Fuze, *The Black People and Whence They Came* (Durban: Killie Campbell Africana Library, 1979), 46. A variation of this story states that Tshaka, doctored by Dingiswayo, deliberately stood over his father casting a shadow on him leading to Senzangakona's illness and death. *JSA*, 4:198, 204, evidence of Ndhlovu.

2. *JSA*, 1:177, 183–84, evidence of Jantshi; *JSA*, 2:185–86, evidence of Mandhlakazi. A similar but slightly different story is found in *JSA*, 2:170, evidence of Makuze. A similar version appears in *JSA*, 3:198–99, evidence of Mkebeni. Likewise, a similar story also exists amongst the Swazi. *JSA*, 2:13, evidence of Mabonsa. Kunene's version is similar to Jantshi's except that it was Zwide's sister who was sent with the aim of discouraging Dingiswayo from attacking Zwide and in order to spy on him. His army thought that he was under the influence of Zwide's muthi when he insisted on visiting Zwide alone. Kunene, *Emperor Shaka*, 116–23. Fuze claims that Dingiswayo visited Zwide on the pretext of courting a woman and was killed. Fuze, *Black People*, 46–47. A. T. Bryant, *Olden Times in Zululand and Natal* (New York: Green, 1929), 158–71. In this text it is Zwide's sister rather than his daughter who secures Dingiswayo's semen, and it is his doctor who "doctors" Dingiswayo. He invades Zwide's land upon learning of his "bewitchment" and the death of his brother-in-law.

3. *JSA*, 1:184–86, evidence of Jantshi; *JSA*, 4:228, evidence of Ndhlovu. Kunene and Fuze state that Tshaka defeated the Ndwandwe by leading them into a trap where they languished for days without food before the Zulu attacked them. Kunene, *Emperor Shaka*, xii; Fuze, *Black People*, 49. Bryant's work emphasizes that the war doctor had sent a plague of rats on the Ndwandwe to eat their food in storage and the bindings of their weapons. Body dirt was collected from inside the chiefdom and sent to Tshaka's doctors.

4. War doctors are generally mentioned, but not seen as a catalyst of change. A notable exception is I. Knight, *The Anatomy of the Zulu Army* (London: Greenhill Books, 1995).

5. These explanations can be found in the oral histories as well as the older written histories of E. Walker, *A History of Southern Africa* (London: 1957); A. T. Bryant, "A Concise History of the Zulu People from the Most Ancient Times," in *Dictionary*; and E. Ritter, *Shaka Zulu* (New York: Green, 1955).

6. Gluckman and Omer-Cooper first posited theories of overpopulation pressure. M. Gluckman, "The Rise of the Zulu Empire," *Scientific American* 202, no. 4 (1969): 157–68. J. D. Omer-Cooper, *The Zulu Aftermath: A Nineteenth-Century Revolution in Bantu Africa* (Evanston: Northwestern University Press, 1966). Guy highlights the demographic and environmental arguments in "Ecological Factors." The impact of trade was first speculated on by Wilson and later expanded by Hamilton and Wright. See M. Wilson and L. Thompson, eds., *The Oxford History of South Africa*, vol. 1, *South Africa to 1870* (New York: Oxford University Press, 1969); C. Hamilton and J. Wright, "Traditions and Transformation: The Phongolo-Mzimkhulu Region in the Late Eighteenth and Early Nineteenth Centuries," in *Natal and Zululand from Earliest Times to 1910*, ed. A. Duminy and B. Guest (Pietermaritzburg: University of Natal, 1989).

7. See J. Cobbing, "The Mfecane as Alibi: Thoughts on Dithakong and Mbolompo," *Journal of African History* 29 (1988): 487–519, and responses to his work by numerous South African historians in Hamilton, ed., *Mfecane*.

8. Callaway's work relied heavily on the testimony of Mpengula Mbande, a member of Callaway's church, as well as his brother, who was a healer in the mid-to late nineteenth century. Reverend Filter transcribed the life story of Paulina Dlamini, another Christian convert who had served under King Cetshwayo between 1925 and 1939. Bryant's work compiled together oral stories of Africans, particularly of the Zulu clan, along with published materials and interviews with white travelers and authorities.

9. T. Mofolo, *Chaka the Zulu* (London: Oxford University Press, 1949); Fuze, *Black People*; Kunene, *Emperor Shaka*.

10. Rosalind Shaw talks about this phenomenon of comparing the power of contemporary politicians to that of witchcraft in R. Shaw, "The Politics and the Diviner: Divination and the Consumption of Power in Sierra Leone," *Journal of Religion in Africa* 26 (February 1996): 30–55.

11. While *insila* referred directly to body dirt, *izidwedwe* meant personal articles with body dirt and generally referred to those things taken from one nation to be used by another. *JSA*, 4:280, evidence of Ndukwana.

12. Ngubane, *Body and Mind*, 33.

13. *JSA*, 1:172, evidence of Hoye.

14. *JSA*, 4:191; Bryant, *Dictionary*.

15. *JSA*, 1:172.

16. This is mentioned even in the late 1970s in Ngubane's *Body and Mind*, 27.

17. Callaway, *Religious System*, 340. *JSA*, 1:172, evidence of Hoye.

18. "It was said that a man who stirred up medicines cast his 'shadow' on the king." *JSA*, 4:280–81, evidence of Ndukwana.

19. Ibid., 281.

20. For Tshaka, see Stuart, *Diary of Fynn*, 31, and Kunene, *Emperor Shaka*, xxiv. For Cetshwayo, see *JSA*, 4:281, evidence of Ndukwana.

21. Callaway, *Religious System*, 341–42, presumably he means *ubulawu*, though he spells it *ubulawo*. Doke, *Zulu-English Dictionary* (1990), also describes isithundu as a "grass vessel for containing a chiefs' charm."

22. *JSA*, 2:170, evidence of Makuza.

23. *JSA*, 4:22, evidence of Mquikana. Isithundu is referred to here by this respondent as medicine that is used to wash the king.

24. Callaway, *Religious System*, 346.

25. Ibid., 343.

26. Ibid., 340; *JSA*, 1:172, evidence of Hoye.

27. *JSA*, 4:280–81, evidence of Ndukwana.

28. Callaway, *Religious System*, 417–18.

29. *JSA*, 3, evidence of Mpatsha ka Sodondo, 296.

30. B. Blair, *Surgical Experiences in the Zulu and Transvaal War, 1879–1881* (Edinburgh, 1883), 10.

31. *JSA*, 3:264, evidence of Mmemi.

32. S. Bourquin, ed., *Paulina Dlamini: Servant of Two Kings* (Pietermaritzburg: University of Natal Press, 1986), 35.

33. Callaway, *Religious System*, 345.

34. From letter of Rev. Owen, dated October 31, 1837: J. Bird, *Annals of Natal 1495 to 1845* (Pietermaritzburg, 1888) vol. 1, 332.

35. *JSA*, 2:250, evidence of Mayinga. *JSA*, 3:287, evidence of Mnkonkoni.

36. *JSA*, 1:25–26, evidence of Baleni.

37. British Parliamentary Papers, C4037, 1884; C. Vijn, *Cetswayo's Dutchman*, 35.

38. J. O. Gump, *The Formation of the Zulu Kingdom in South Africa, 1750–1840* (New York: Mellen Press, 1990), 113–16.

39. For details of this ceremony see Knight, *Anatomy*, 123–52.

40. First half of this description found in *JSA*, 1:40, evidence of Baleni, specific plants related in Callaway, *Religious System*, 345.

41. Use of herbs from Hutchings, *Zulu Medicinal Plants. Umvithi* historical use recorded among the Sotho, 20; *umabophe*, 159; *umatshwilitshwili*, 229; *imfingo*, 12.

42. *JSA*, 4:373, evidence of Ndukwana.

43. Bourquin, *Paulina Dlamini*, 69–70.

44. Ibid., 37.

45. Ibid., 34–35.

46. *JSA*, 1:40–41, evidence of Baleni.

47. J. Guy, "Production and Exchange in the Zulu Kingdom," in *Before and After Shaka*, ed. J. Pieres, 45.

48. This is most likely the reason he would allow his female bodyguards to use guns. *JSA*, 3:328, evidence of Mpatshana; Bourquin, *Paulina Dlamini*, 54.

49. *JSA*, 4:63–64, evidence of Mtshapi.

50. *JSA*, 4:26, evidence of Mqaikana.

51. Stuart, *Diary of Fynn*, 282.

52. *JSA*, 3:312, evidence of Mpatshana.

53. *Native Laws and Customs Commission*, September 1881, 6–7, evidence of Shepstone.

54. C. de B. Webb and J. B. Wright, *A Zulu King Speaks: Statements Made by Cetshwayo kaMpande on the History and Customs of His People* (Pietermaritzburg: University of Natal Press, 1987), 90.

55. Ibid., 90–99. Also see *JSA*, 4:11, evidence of Ndukwana.

56. Death had to be reported to the chief or king. "Reports of Native Affairs, Inanda," 1875 *Blue Book on Native Affairs*.

57. *JSA*, 4:319–20, evidence of Ndukwana.

58. *JSA*, 1:26, evidence of Baleni, Madhlodhlongwana's position is not identified within the text.

59. Ibid., 26.

60. Ibid., 25–27.

61. *JSA*, 1:313–15.

62. *JSA*, 3:181.

63. Ibid.

64. Mpande: *JSA*, 3:181; C. Webb and J. Wright, eds., *A Zulu King Speaks*, 19; Samuelson, *Long, Long Ago*, 35, 229.

65. Quote from *JSA*, 2:130. Also see *JSA*, 1:19, 314–15; *JSA*, 3:181.

66. *JSA*, 1:19, evidence of Baleni.

67. *JSA*, 1:313, evidence of Lunguza.

68. *Native Laws and Customs Commission 1881* (Grahamstown, 1881), 68, evidence of Shepstone. He first mentions their role in a memo from 1864 in "Questions by Lieut Governor Scott, on the Conditions of the Natives in Natal and Answers by the Secretary for Native Affairs," *Blue Book* (London, 1875), 7.

69. *JSA*, 4:319–20; *JSA*, 1:25–27, evidence of Baleni.

70. *JSA*, 4:380, evidence of Ndukwana.

71. Lady Barker, *A Year's Housekeeping in South Africa* (1877; repr., London: Macmillan, 1883), 188.

72. *JSA*, 1:192.

73. All quotes and information from *JSA*, 1:342, evidence of Lunguza. Unfortunately a time period is not given for Botshobana's actions.

74. *JSA*, 2:250–51, evidence of Mayinga.

75. *JSA*, 1:90, evidence of Kambi.

76. Callaway, *Religious System*, 195. Evidence suggests that this was during the time of Mpande, though it does not explicitly state this as fact.

77. *JSA*, 1:195, evidence of Jantshi; and *JSA*, 1:9, 320; *JSA*, 5:40.

78. *JSA*, 4:317, evidence of Ndukwana.

79. *JSA*, 1:9–10, evidence of Baleka.

80. To Secretary for Native Affairs (SNA) Pretoria from Arthur Shepstone, November 20, 1911, SAB, NTS, 9304, 10/376.

81. Mpande: *JSA*, 4:380; Cetshwayo: *JSA*, 3:315, evidence of Mpatshana; also Webb and Wright, *A Zulu King Speaks*, 20.

82. *JSA*, 3:315.

83. Ibid. Mpatshana states that military service was akin to starvation and being beaten.

84. Callaway, *Religious System*, 340n86.

85. Ibid., 345.

86. Ibid., 341. Also see *JSA*, 3:240, evidence of Mmemi.

87. J. Laband, *Rope of Sand: The Rise and Fall of the Zulu Kingdom in the Nineteenth Century* (Johannesburg: Jonathan Ball, 1995), 27.

88. Makanna and Mlanjeni of the Xhosa, Nehanda and Kagubi of the Shona; such healer-led rebellions happened throughout Africa. See I. Berger, "Rebels or Status-Seekers? Women as Spirit Mediums in East Africa," in *Women in Africa: Studies in Social and Economic Change*, ed. N. Hafkin and E. Bay (Stanford: Stanford University Press, 1976); S. Feierman, "Healing as Social Criticism in the Time of Colonial Conquest," *African Studies* 54 (1995): 72–85. This move from public to private healing also occurred among the Ku people of Liberia. See E. Tonkin, "Consulting Ku Jlople: Some Histories of Oracles in West Africa," *Journal of the Royal Anthropological Institute* 10 (2004): 539–60.

89. For an analysis of the 1906 uprising and the role of both inyangas and human muthi see J. Guy, *The Maphumulo Uprising* (Pietermaritzburg: University of KwaZulu-Natal Press, 2005). Guy argues that healers did not play as important a role as colonialists claimed though they were clearly employed in the event.

90. C. Jackson, "Native Superstition in Its Relation to Crime," *South African Association for Advancement of Science* (1917): 251–60; also see Rev. J. Tyler, *Forty Years among the Zulus* (Boston, 1891), 97; Lugg, *Natal Family*, appendix, 8–9.

91. Stuart, *Diary of Fynn*, 11.

92. Lugg, *Natal Family*, 96–97, and Knight, *Anatomy*, 171–72. These two stories are somewhat different in terms of who was killed and how.

93. *JSA*, 3:327.

94. Swazi: *JSA*, 4:14, 298; Zibebu: *JSA*, 3:327. Knight contends that this took place during this time (*Anatomy*, 170).

95. *JSA*, 2:55–56.

96. An early case in Zululand dates from 1887. "Minute Paper," PAR 1/MEL/ 3/2/1 refers to a body considerably mutilated. For the later period see C. Jackson, "Native Superstition in Its Relation to Crime," *South African Association for Advancement of Science* (1917): 251–60; PJS, "A Zulu Boy's Heart," *The Nongqai* (March 1921); W. L. Speight, "Human Sacrifice in South Africa," *The Nongqai* (February 1935).

97. Turrell's work is based on an analysis of court records from 1900 to 1930 that look at muthi killings. R. Turrell, "Muti Ritual Murder in Natal: From Chiefs to Commoners (1900–1930)," *South African Historical Journal* 44 (2001): 21–39.

CHAPTER 3: AFRICAN-WHITE ENCOUNTERS

1. An account of this party is in Lady Barker, *A Year's Housekeeping*, 169–86. While the majority of travelers' accounts were produced by men, I have chosen this particular example because it is rather rich in detail and representative of colonial perspectives and colonists' fascination with healers.

2. Barker, *A Year's Housekeeping*, 171, 172, 186.

3. Pamphlet: "Your Health: A Guide to the Medicine and Public Health Building at the New York World's Fair 1940, Containing the Answers to Many Fundamental Questions on Health by the American Museum of Health, Inc." (New York, 1940).

4. J. Pieterse, *White on Black: Images of Africa and Blacks in Western Popular Culture* (New Haven, CT: Yale University Press, 1992), 69–75. Vaughan, *Curing Their Ills*, chapters 3 and 7.

5. See Etherington, "Missionary Doctors," 77–91.

6. M. Vaughan, *Curing Their Ills*; T. Ranger, "Godly Medicine: The Ambiguities of Medical Mission in Southeastern Tanzania," in *Social Basis*, ed. Feierman and Janzen; J. Comaroff, "The Diseased Heart of Africa: Medicine, Colonialism, and the Black Body," in *Knowledge, Power, and Practice: The Anthropology of Medicine and Everyday Life*, ed. S. Lindenbaum and M. Lock (Berkeley and Los Angeles: University of California Press, 1993), 305–29; P. Landau, "Explaining Surgical Evangelism in Colonial Southern Africa: Teeth, Pain, and Faith," *Journal of African History* 37, no. 2 (1996): 261–81; A. Digby and H. Sweet, "Nurses as Culture-Broker in Twentieth-Century South Africa," in *Plural Medicine, Tradition, and Modernity, 1800–2000*, ed. W. Ernst (New York: Routledge, 2002); Zondi, "African Demand and Missionary Charity: The Development of Mission Health Services in Kwazulu to 1919," University of KwaZulu-Natal African Studies Seminar: www.history.und.ac.za/Sempapers/Zondi.PDF (accessed January 25, 2006).

7. D. Carlson, *African Fever: A Study of British Science, Technology, and Politics in West Africa, 1787–1864* (New York: Science History Publications, 1984), 5. What began as an exchange focused on individual treatment had expanded by the 1850s to include bioprospecting of indigenous flora for the local medical market, and by the twentieth century involved the export of indigenous South African remedies overseas.

8. Zondi, "African Demand."

9. Stuart, *Diary of Fynn*, 85.

10. Hale, *Norwegian Missionaries*, 21, 35.

11. King: S. Gray, ed., *The Natal Papers of John Ross* (Durban: Killie Campbell Africana Library Press, 1992), 122; Peterson: Stuart, *Diary of Fynn*, 4.

12. Stuart, *Diary of Fynn*, 4.

13. Conceit quote: Stuart, *Diary of Fynn*, 277; after recovering from fever and noting the African medicines used, he insists that the same treatment be given to his friend. Stuart, *Diary of Fynn*, 43.

14. *JSA*, 1:60.

15. S. G. Campbell, "Zulu Witchdoctors," 10.

16. Hale, *Norwegian Missionaries*, 36. Champion also reports that many people asked him for medicines: Booth, *Journal of Rev. George Champion*, 80. During the attack on Port Durban in the 1830s, Rev. Owens reports on the treatment of wounded Zulus by Mrs. Champion. G. Cory, *The Diary of the Rev. Francis Owen* (Cape Town: Van Riebeeck Society, 1926), 159; Astrup boasts in a letter home in 1899 that people come to him for medicine and to have teeth extracted. Hale, *Norwegian Missionaries*, 153.

17. A. Digby, "Self-Medication and the Trade in Medicine within a Multi-ethnic Context: A Case Study of South Africa from the Mid-nineteenth to the Mid-twentieth Centuries," *Social History of Medicine* 18, no. 3 (2005): 439–57.

18. *JSA*, 2:29.

19. *Government Gazzette* 14, no. 689 (January 14, 1862); S. G. Campbell, "Zulu Witchdoctors," 12.

20. "Natal Medical Council," *The Probe* 1, no. 10 (August 1911); W. H. Haupt, "Letter to the editor, Native Doctors," *The Journal of the Medical Association of South Africa* 4, no. 22 (November 22, 1930).

21. S. G. Campbell, "Zulu Witchdoctors," 10.

22. Colony of Natal, Department Reports 1898, Nqutu, bb16.

23. Hale, *Norwegian Missionaries*, 36, 50, 153, 155; based on a careful reading of *Natal Blue Books* from 1874 to 1910. The topic of competition between various practitioners will be discussed in more detail within the next chapter.

24. Fear of amputation: "Hygiene of the Zulus," *The South African Medical Journal* (Transvaal), March 8, 1889; *JSA*, 3:227; H. Kuper, "Medicine among the South Eastern Bantu," *The Leech* (November 1936); P. Keen, "The Witch Doctor," *The Leech* (June 1948). The 1895 smallpox outbreak resulted in a quarantining of the Nhlazatshe area, the missionary doctor complains that such rules were ignored and Africans snuck out in the dead of night. Hale, *Norwegian Missionaries*, 160. For fear of quarantining see *Blue Book of Native Affairs*, 1903 Ndwedwe, *Blue Book of Native Affairs*, 1905 Ndwedwe and *Natal Blue Book*, 1896, "Departmental Reports," B37. For fear of hospitals see comments of District Surgeons in local Blue Books: e.g., 1875 *Blue Book on Native Affairs*; 1881 *Blue Book on Native Affairs*, 72; *Natal Blue Book*, 1894–95, B63.

25. C. Ballard, "Traders, Trekkers, and Colonists," in *Natal and Zululand*, ed. Duminy and Guest, 118.

26. This is obviously not reflective of the original name, as Zulu contains very few *R*'s, and it is evident from Isaacs's text that he often misspells Zulu words and names and adds *R*'s.

27. N. Isaacs, *Travels and Adventures*, 243.

28. Ibid.

29. As one of Stuart's respondents explains, "It used formerly to be the practice for women and girls to expose themselves before cowards by way of insulting them, and say as they did so, 'Is this what you want? It's all that you're good for.'" *JSA*, 1:124.

30. Ibid., 234.

31. Ibid., 295.

32. For a discussion on the repercussions of criminalizing healers see G. St. J. Orde Browne, "Witchcraft and British Colonial Law," *Africa, Journal of the International Institute of African Languages and Cultures* 8, no. 4 (1935): 481–87. C. Clifton Roberts, "Witchcraft and Colonial Legislation," *Africa* 8, no. 4 (1935): 488–93. F. Melland, "Ethical and Political Aspects of African Witchcraft," *Africa* 8, no. 4 (1935): 495–503.

33. Ballard, "Traders," 125. More recent work on the colonial period in Ghana shows that the British moved from an outright banning of witchfinders to a process that enabled "self-confessed" witches to seek treatment from African healers. See N. Gray, "Independent Spirits: The Politics of Policing Anti-Witchcraft Movements in Colonial Ghana, 1908–1927," *Journal of Religion in Africa* 35, no. 2 (2005): 139–58.

34. B. Kline, *Genesis of Apartheid: British African Policy in the Colony of Natal, 1845–1893* (Lanham: University Press of America, 1988), 2.

35. Ibid.

36. Ibid., 11.

37. Ballard, "Traders," 125.

38. Kline, *Genesis of Apartheid*, 6.

39. N. Etherington, "The 'Shepstone System' in the Colony of Natal and beyond the Borders," in Duminy and Guest, *Natal and Zululand*, 178.

40. Etherington, "Shepstone System," 171–72.

41. E. Brookes, *White Rule in South Africa 1830–1910: Varieties in Governmental Policies Affecting Africans* (Pietermaritzburg: University of Natal Press, 1974), 135.

42. Etherington, "Shepstone System," 173.

43. Ibid., 172–73.

44. Ibid. This issue was brought up at a meeting protesting the system in 1863.

45. D. Welsh, *The Roots of Segregation: Native Policy in Colonial Natal, 1845–1910* (Cape Town: Oxford University Press, 1975), 16.

46. PAR, 1/BLR, 6, 1878, 37. "Minute Paper containing notes on a case in which a native was charged for practicing as a witchdoctor."

47. "Annexure to the Crown Prosecutor's Report, December 26, 1851," *South African Archival Records* (Cape Town, 1962), 312–20.

48. E. Brookes, *White Rule*, 52.

49. *South African Archival Records: Natal*, no. 3 (Cape Town, 1963), see pages 312–32.

50. Ibid., 321. My emphasis.

51. "Reports of Native Affairs, Umgeni," *Blue Book on Native Affairs*, 1875; and E. Neville, "Report of the Government Chemist," *Government Blue Book*, vol. 2, 1899, H28. For a breakdown of these cases see K. Flint, "Negotiating a Hybrid Medical Culture," appendix 2.

52. *Natal Native Affairs Commission 1852*, 10, evidence of Struben.

53. Brookes, *White Rule*, 52.

54. *South African Archival Records: Natal*, no. 3, 320. Emphasis in the original.

55. Ibid., 321–22.

56. Ibid., 322.

57. See C. Crais, *White Supremacy and Black Resistance in Pre-industrial South Africa: The Making of the Colonial Order in the Eastern Cape, 1770–1865* (Johannesburg: Witwatersrand University Press, 1992); J. Pieres, *The Dead Will Arise* (Johannesburg: Ravan Press, 1989), 124–38. Mlanjeni's practice of finding witches is much like that of other witch-finding movements that occurred through out Africa during the colonial period. See A. Richards, "A Modern Movement of Witch-Finders," *Africa* 8, no. 4 (1935): 448–60.

58. Pieres, *The Dead Will Arise*, 55.

59. *Natal Native Commission 1852*, 10, evidence of Resident Magistrate Struben. He makes the radical recommendation that "the doctor who is usually employed to discover the wizard would be made to prove his guilt, beyond his mere assertion of it, before the Resident Magistrate and not before the chief." Otherwise Struben recommends fining doctors who make accusations of witchcraft "without proof other than his own declaration."

60. *British Parliamentary Papers: Report from the Select Committee on the Kafir Tribes with Proceeding Minutes of Evidence* (House of Commons, 1851; reprint, 1968), 17.

61. *Natal Native Commission 1852*, 25, evidence of Shepstone.

62. Ibid., 11, evidence of Boschoff.

63. Ibid., 64, evidence of David.

64. Ibid., 59, evidence of Fynn.

65. "Questions by Lieutenant Governor Scott, on the Conditions of the Native in Natal and Answers by the Secretary for Native Affairs," October 16, 1864, *Natal Blue Book 1866* (London, 1866), 7.

66. *Native Laws and Customs Commission 1881* (Grahamstown, 1881), 68, evidence of Shepstone.

67. T. Shepstone, "Letter to the editor," *Natal Mercury*, January 19, 1892.

68. Etherington, "Shepstone System," 19, 179. For hereditary chiefs also see M. Mahoney, "Between the Zulu King and the Great White Chief: Political Culture in a Natal Chiefdom, 1879–1906" (PhD diss., University of California, Los Angeles, 1998), 65–74.

69. PAR, SNA, I/1/35, 1878/774.

70. "Return of Cases in which Rain Doctors and Witchdoctors Have Been Punished for the Exercise of Their Craft," Minute for Weenen Magistrate, PAR, SNA, I/1/30, 1878/936.

71. "Report of the Acting Secretary for Native Affairs on the Subject of the Applicants to the so called Rainmakers, also to male and female Witch Doctors," June 24, 1878; PAR, SNA, I/1/35, 1878/774.

72. Ibid. This categorization seems odd on a report that outlines punishments for witchdoctors and raindoctors, and one wonders if the government was still attempting to intervene in and discern cases of witchcraft.

73. *1881 Natal Laws and Customs Commission*, 521, evidence of Cetshwayo.

74. This is a bit conjectural given that nothing in any secondary sources or the primary sources explicitly states why the government wanted to prevent healers from going into Zululand.

75. "Report of the Acting Secretary for Native Affairs on the Subject of the Applicants to the so called Rainmakers, also to male and female Witch Doctors," June 24, 1878; PAR, SNA, I/1/35, 1878/774.

76. Brookes, *White Rule*, 144–45; *South African Criminal Law and Procedure* (Cape Town: Juta, 1946), 1499.

77. Zululand Proclamation 11. *South African Criminal Law and Procedure*, 1499.

78. Grey to Smith, December 10, 1847, no. 416, PRO CO 179/2, as quoted in Kline, *Genesis of Apartheid*, 11.

79. C. Saunders, *Historical Dictionary of South Africa* (Lanham, MD: Scarecrow Press, 1983).

80. N. Etherington, *Preachers, Peasants, and Politics in Southeast Africa, 1835–1880* (London: Royal Historical Society, 1978), 47–53.

81. Hale, *Norwegian Missionaries*, 85. Likewise Champion noted that "those in power look upon us as introducers of some greater object of reverence than Dingana and suspect that those who come to hear our words will be enticed away from their king." Booth, *Journal of Rev. George Champion*, 117.

82. M. Vaughan, "The Great Dispensary in the Sky," in *Curing Their Ills*, 55–77; P. Landau, "Explaining Surgical Evangelism in Colonial Southern Africa: Teeth, Pain, and Faith," *Journal of African History* 37, no. 2 (1996): 261–81.

83. Etherington, *Preachers*, 103.

84. Samuelson, *Long, Long Ago*, 30–32.

85. *Native Commission 1903–05*, vol. 5, 126, evidence of J. Crosby, Zululand trader. *1882 Natal Blue Books*, evidence under Umsinga.

86. Reyher, *Zulu Woman*, 28–32.

87. Hale, *Norwegian Missionaries*, 167.

88. *JSA*, 5:201.

89. *JSA*, 2:185, evidence of Mandhlakazi.

90. Hale, *Norwegian Missionaries*, 1851, correspondence of Scheuder, February, 35.

91. Veritas, *The Kafir Labour Question, as it affects the Colony of Natal, being an exposition of the Existing policy of Government shewing the evils and suggesting remedies* (Durban, 1851), 17.

92. Tyler, *Forty Years*, 100.

93. Quoted in J. McCord, "Zulu Witch Doctor," 307.

94. Ibid.

95. Etherington, *Preachers*, 136; Veritas, "The Kafir Labour Question," 17.

96. *JSA*, 4:14; *JSA*, 5:139.

97. Etherington, *Preachers*, 136.

98. In 1873 in Pietermaritzburg, any African exceeding five days in town was obliged to accept any job offered at the standardized monthly rate. Ibid.

99. H. Slater, "The Changing Pattern of Economic Relationships in Rural Natal, 1838–1914," *Economy and Society in Pre-Industrial South Africa*, ed. S. Marks and A. Atmore (London: Longman, 1980), 158. In hopes of creating a cash need, the government placed heavy tariffs on African-purchased imports, mandated European-style dress in the cities, required rent of Africans squatting on white farms or Crown land areas, and imposed taxes while remitting the taxes of persons serving colonial farmers. Etherington, *Preachers*, 12.

100. While this is difficult to assess, as governmental records indicate government's interest in the prosecution of cases not actual numbers, this was the perception among the African population and mirrors the experience of Africans in other British colonies where anthropologists reported an increase in witch-finder movements and use of protection charms against witchcraft. See A. Richards, "A Modern Movement of Witch-Finders," *Africa* 8, no. 4 (1935): 448–60. F. Melland, "Ethical and Political Aspects of African Witchcraft," *Africa* 8, no. 4 (1935): 495–503. For neighboring Pondoland see M. Hunter, *Reaction to Conquest* (London: Oxford University Press, 1936), 43.

101. This was likewise the case in other British colonies in Africa. See C. C. Roberts, "Witchcraft and Colonial Legislation" *Africa* 8, no. 4 (1935): 488–94; Melland, "Ethical and Political Aspects of African Witchcraft," 495–503.

102. *Natal Native Commission 1852*, 89, evidence of Umfulatela.

103. *Natal Native Commission 1881*, 355, evidence of Joko speaking on behalf of the young Chief Silwane of the Amacune.

104. *Evidence taken before the Natal Native Commission 1881*, 336.

105. *JSA*, 3:182, evidence of Mkando.

106. *Evidence taken before the Natal Native Commision 1881*, 336.

107. Ibid., 350.

108. *JSA*, 1:69.

109. *Natal Native Commission 1881*, 212–13, 251.

110. *Natal Native Commission 1881*, 363, evidence of Tyutyella of the Makanya.

111. *JSA*, 1:137, evidence of Gama.

112. Rev. H. Callaway, Notes for *Nursery Tales, Traditions and Histories of the Zulus* (1868; repr., Westport, CT: Negro Universities Press, 1970), 351. Tikoloshe or tokoloshe.

113. Tyler publishes this in 1891, but notes that he learns this from S. C. Samuelson (*Forty Years*, 112); Kwa Ndlulamiti pamphlet selling tokoloshe fat in 1920, PAR,

NTS, 9303, 8/376, complaints about its being sold as a medical fat in Natal appear in "Minutes of United Transkien Territories General Council Session." PAR, CNC, 50A, 1934; "How the Tokolotshe Takes his Victims" (1937) *Natal Mercury*, KCL clippings book, no. 10, 71; H. Ngubane, who got much of her information in the 1970s from oral interviews writes that this apparition was introduced by the Mpondo, a group that borders the Xhosa and the Zulu during the period of migrant labor (*Body and Mind*, 34).

114. Callaway, *Religious System*, 330–32; "Zulu Customs; Kafir Doctors," *Natal Colonist*, November 11, 1879. Later authors also suggest that the origins of bone throwing forms of divination amongst the Zulu came from the Basotho. Bryant claims stick throwing was from the Tongas. Bryant, "Cult," 142. S. Berglund includes a discussion of these origins on 194n64. Photo from Killie Campbell Collection, B. Kish, *Photographs of Natal and Zululand* (Durban, 1882).

115. Female healers are evident in the record from the time of Tshaka and throughout the Zulu kingdom. Fynn tells a rather suspect story about Tshaka killing female healers, but at least this points to their existence. Stuart, *Diary of Fynn*, 155–56; *Natal Native Commission* 1852, vol. 5, 64, evidence of Fynn. While one of Stuart's respondents claims that all of Tshaka and Dingane's royal isangomas were men, another disagrees referencing the assegais and shields that women isangomas used. *JSA*, 1:355, 48. Fynn writes that healers who practiced divination could be either men or women who entered the profession at the insistence of the ancestors, and learned the art from an experienced renowned healer. The *inyanga yemithi* occupation, however, was usually inherited father to son. Stuart, *Diary of Fynn*, 280. While the majority of inyanga families passed specialized herbal knowledge through the male line, Fynn claimed that women became privy to such knowledges when no sons could carry on the family tradition or when the ancestors indicated otherwise. 1852 *Natal Native Commission* vol. 4, 55, evidence of Fynn.

116. Mpatshana claimed that six out of ten men were isangomas during this time period, *JSA*, 3:315; other mentions of these regiments: *JSA*, 4:380, and *1881 Native Laws and Customs Commission*, 521.

117. Bryant claimed in 1917 that 90 percent of isangomas were women. This is the only source that attempts to give a percentage to the sex of these healers ("Cult," 142). Travel narratives from the late ninteenth century often mention female healers as opposed to male ones. S. Hanretta challenged this notion, hypothesizing that women frustrated by their exclusion in the new Zulu state and by state impositions in their personal lives became diviners in the mid-nineteenth century. S. Hanretta, "Women, Marginality and the Zulu State: Women's Institutions and Power in the Early Nineteenth Century," *Journal of African History* 39, no. 3 (1998): 389–415. The historical record, however, does not support this hypothesis as it is nearly impossible to tell the gender of isangomas in the Tshakan era and difficult to tell the exact gender composition in the mid-nineteenth century. Furthermore his own evidence is taken from Bryant, who points to a much later date for this transformation.

118. Shooter, *The Kafirs of Natal*, 192; Callaway, *Religious System*, 266.

119. J. Gussler, "Social Change, Ecology, and Spirit Possession among the South African Nguni," in *Religion, Altered States of Consciousness, and Social Change*, ed. E. Bourguignon (Columbus: Ohio State University Press, 1973), 88–126.

120. *JSA*, 3:171, evidence of Mkando.

121. B. Sundkler, *Bantu Prophets in South Africa* (London: 1961), 48.

122. Ibid., 109.

123. For a full discussion see ibid., 256–64, and appendix A, 350.

124. Ibid., 213–15.

125. *1881 Natal Native Commission*, 14. As stated "for weaning people from trusting in any who pretend to practice witchcraft, or profess to have supernatural power." Fitzgerald quote from M. Gelfand and P. W. Laidler, *South Africa: Its Medical History, 1652–1898* (Cape Town: C. Struik, 1971), 299.

126. *1881 Natal Native Commission*, 18, 33, 60, 68, 131, 164, 175.

127. *Report of the Indian Immigrants Commission* (Pietermartizburg, 1882).

128. The debate surrounding the training of African biomedical doctors in the 1910s and '20s also carried this rhetoric. I. S. Monamodi, "Medical Doctors under Segregation and Apartheid: A Sociological Analysis of Professionalization among Doctors in South Africa, 1900–1980" (PhD diss., Indiana University, 1996).

129. See *Select Committee on the Contagious Diseases Prevention Bill*, no. 19, 1890, SSP-LC.

130. Monamodi, "Medical Doctors," 65; K. Shapiro, "Doctors or Medical Aids—The Debate over the Training of Black Medical Personnel for the Rural Black Population in South Africa in the 1920s and 1930s," *Journal of Southern African Studies* 13, no. 2 (1987): 234–55.

131. Suggestion to license and register healers made at the *1881 Natal Native Commission*, 351, evidence of Ralfe; "Memorandum by JCC Chadwick, RM Ixopo for the consideration of the Honourable Secretary for Native Affairs on the subject of alterations of the Code of Native Law," PAR, SNA, 1/6/18.

132. Barker, *A Year's Housekeeping*, 185.

133. To name but a few: C. Barter, *Alone among the Zulus* (Pietermaritzburg: University of Natal Press, 1995), 43–46; E. G. C., "Zulu Customs: An Interview with an Isanusi, or Kafir Witch-Doctor," *Natal Colonist*, June 19, 1879, Reprint at the South African Library, Cape Town; Story of Rev. Adams as told in T. Jenkinson, *Amazulu: The Zulus, Their Past History, Manners, Customs, and Language: Natal 1873–1879* (London, 1882), 109–11.

134. Barker, *A Year's Housekeeping*, 185.

135. An excellent source for early shows in England is R. Altick, *The Shows of London* (Cambridge, MA: Harvard University Press, 1978).

136. Barker, *A Year's Housekeeping*, 178.

137. All information for this showing to be found in Altick, *The Shows of London*, 282–83.

138. Tyler, *Forty Years*, 100.

139. Barker, *A Year's Housekeeping*, 179.

140. E. G. C., "Zulu Customs: An Interview with an Isanusi, or Kafir Witch-Doctor."

141. Barker, *A Year's Housekeeping*, 172 (emphasis in the original).

142. F. Fanon, *Wretched of the Earth* (New York: 1963), 38–40.

143. A. Stoler, *Race and the Education of Desire* (Durham, NC: Duke University Press, 1995).

144. Guy, *The Maphumulo Uprising*.

145. The issues of hill stations is addressed in a number of chapters, but particularly Ann Stoler's in J. Breman, ed., *Imperial Monkey Business: Racial Supremacy in Social Darwinist Theory and Colonial Practice* (Amsterdam: VU University Press, 1990).

146. Green, *Indigenous Theories*, 11–12.

CHAPTER 4: COMPETITION, RACE, AND PROFESSIONALIZATION

1. Quote from "Natives Have Own 'Medical' Body," *Sunday Times* (Johannesburg), April 10, 1938; "Native Medical Association," *Star*, April 11, 1938.

2. Ngcobo to Minister of Native Affairs, Cape Town, March 2, 1931; PAR, NTS, 9301, 1/376, 3.

3. *JSA*, 3:179, evidence of Mkando.

4. Quote from *Natal Native Affairs Commission 1906–07 Evidence*, 725. Also see 831, 871, 884–86, 895.

5. Ibid., 884.

6. Ibid., 832.

7. Ibid., 772.

8. Ibid., 814.

9. Ibid., 792.

10. Ibid., 772.

11. Ibid., 832.

12. PAR, NTS, 2/376, pt. 1 (Magistrate Verulam to SNA Pretorian Feb 24, 1930). Several chiefs or magistrates, when recommending persons for inyanga licenses, mention that the person is no longer fit for manual labor (e.g., application for Izinyanga license, Hlomis Dhlamini, April 8, 1936, PAR, GES, 1785).

13. First mention of this is 1894 letter to editor of the *South African Medical Journal:* "The Grievances of District Surgeons, Letter to the Editor," *South African Medical Journal* 2, part 2 (June 1894), 63.

14. This notion is uniformly supported by oral testimony from healers today, as well as in the archival sources (see letter to "SNA Pretoria from Chief Native Commissioner [CNC] Natal," "Native Medicine Men," October 6, 1915, PAR, CNC, 193, 1915/149). Foreign healers whose reputations preceded them were invited and compensated by Zulu kings and chiefs to live in their compounds.

15. Complaint of nonlicensed traveling inyangas: *Natal Native Affairs Commission 1906–7 Evidence*, 828; complaint "from CNC Natal to SNA" Pretoria,

October 6, 1915, PAR, NTS, 9301, 1/376, 1; "Natives Have Own 'Medical' Body." Examples from Ubombo and Nongoma "HA Rippon (Sergt. SAMR) to Magistrate of Umbombo," July 26, 1918, and "Oswald Fynney (magistrate of Nongoma) to CNC Natal," September 4, 1918. Both cases regard a traveling healer from up north who administered snuff to prevent illness from malaria that was epidemic at the time. SAB, NTS, 9465, 3/394.

16. "CNC Natal to SNA Pretoria," October 6, 1915, PAR, NTS, 9301, 1/376, 1; "Digest of Replies to CNC Circular 4/15, re Sale of Roots, etc. by Native Medicine Men," 1916, PAR, NTS, 9301, 1/376, 1.

17. L. Kuper, H. Watts, and R. Davies, *Durban: A Study in Racial Ecology* (London: J. Cape, 1958), 53. See chart for population growth based on race from 1862 to 1951, based on a compilation of census reports. In 1880 only four thousand Africans lived in Durban; this increased to over twelve thousand by 1920.

18. "CNC Natal to SNA Pretoria," October 6, 1915, PAR, NTS, 9301, 1/376, 1. P. Cele tells how his father used to collect herbs around Durban as well as buy herbs from women coming into the city during the 1930s. P. Cele, interview, December 9, 1998, Durban.

19. P. Cele interview.

20. A. B. Cunningham, *An Investigation of the Herbal Medicine Trade in Natal/KwaZulu* (Pietermaritzburg: Institute of Natural Resources, University of Natal, 1988); J. Beall and E. Reston-Whyte, "How Does the Present Inform the Past? Historical and Anthropological Perspectives on Informal Sector Activity in Early Twentieth Century Natal," paper presented at the Conference on the History of Natal and Zululand, University of Natal, Durban, July 1985.

21. "1911 Report re Native Doctors and sale of Medicines in Native Eating Houses, Durban," PAR, NMC, T23, 1208/1911.

22. Approximately 21 percent of Zulu, Xhosa, and Sotho herbs are poisonous; they are used to induce vomiting. A. Hutchings, "Acute Poisoning in Zulu and Xhosa Traditional Medicine," paper presented at the Premier colloque européen d'ethnopharmacologie, March 23–25, 1990, 123–32.

23. P. Cele interview.

24. *Braby's Natal Directory* (Braby Publisher), years 1930–39, lists "native chemists."

25. "Natives Have Own 'Medical' Body," *Sunday Times*, April 10, 1938; "Native Medical Association"; Natal Medical Council to Barret, February 18, 1921, PAR, NTS, 9301, 1/376, 2. Gordon to Secretary of Natal Medical Council, December 15, 1924, PAR, NTS, 9301, 1/376, 2.

26. This comes from Ngcobo's 1931 letterhead, PAR, CNC, 50A.

27. I. Alexander's pamphlet (1930), SAB,GES, 1788, 25/30M. Quote from I. Alexander's letterhead on letter dated 1939, PAR, 1/DBN 1/1/1/2/2.

28. This is probably downplayed within the historiography as the editors of the *South African Medical Record* did not include African patients in their analysis of competition. Monamodi, however, rightly recognizes that African healers were a significant competitive threat to whites doctors ("Medical Doctors," 76). Shapiro

also acknowledges that whites doctors feared the competition that full medical training might bring if it were available to African doctors ("Doctors or Medical Aides.")

29. *South African Medical Record* 1, no. 2 (1903): 32; For a fuller discussion see A. Digby, "'A Medical El Dorado'? Colonial Medical Incomes and Practise at the Cape," *Social History of Medicine* 7, no. 3 (1995): 463–79; Monamodi, "Medical Doctors," 122–216.

30. These conclusions are derived from a close reading of the Natal Blue Books from the 1870s to 1910. Evidence of new compliance to small pox vaccine, 1898 *Natal Blue Book*, evidence from Umgeni; 1904 *Blue Book of Native Affairs*, evidence from Umvoti.

31. For a history of the emergence of public health in South Africa see S. Marks and N. Andersson, "Typhus and Social Control: South Africa, 1917 to 1950" in *Disease, Medicine, and Empire: Perspectives on Western Medicine and the Experience of European Expansion*, ed. R. MacLeod and M. Lewis (New York: Routledge, 1988); S. Parnell, "Creating Racial Privilege: The Origins of South African Public Health and Town Planning Legislation," *Journal of Southern African Studies* 19, no. 3 (September 1993): 471–88; A. MacKinnon, "'Of Oxford Bags and Twirling Canes': The State and Popular Responses to Anti Malarial Campaigns in the Early Twentieth Century Zululand," *Radical History Review* 80 (Spring 2001): 75–99.

32. M. McCord, *Calling*, 173–74. Also see "The Grievances of District Surgeons, letter to the Editor," *South African Medical Journal* 2, part 2 (June 1894): 63; and J. McCord, *Patients*, 89, 201.

33. "To Secretary for Public Health, Pretoria from Magistrate Escourt," July 11, 1929, PAR, CNC, 50A.

34. J. McCord, *Patients*, 274.

35. M. McCord, *Calling*, 173.

36. "To Secretary Natal Medical Council," November 1, 1914, PAR, NMC, T24.

37. Ibid., 224.

38. "REX v. Mafavuke Ngcobo" (1940), SAB, GES, 1788, 25/30M, 174, 234.

39. Campell, "Zulu Witchdoctors."

40. J. McCord, *Patients*, 119–21.

41. To the Deputy Chief Health Officer from Under Secretary for Public Health, November 24, 1937, SAB, GES, 1786, 25/30H.

42. "Secretary for the interior to SNA," June 28, 1911, PAR, NTS, 9301, 1/376, 1.

43. For detailed information on Afrikaner home remedies see A. Coetzee, "Some Folkloristic Aspects of Afrikaans Folk Medicine," *Institute for the Study of Man* 5 (1962): 1–23. Also see *Report of the Carnegie Commission, The Poor Whites Problem in South Africa* (Pretorian, 1932).

44. "IBM to Clark," November 13, 1920, PAR, NTS, 9301, 1/376, 1.

45. "Skies Split!" *South African Post*, September 21, 1947.

46. I. Alexander's pamphlet, 1930, PAR, GES, 1788, 25/30M.

47. "Reasons for Judgement in the Case of Mafayifa Radebe," PAR, NMC, T23, 5057/1905.

48. "Dr. Campbell Watt to Colonial Secretary," August 15, 1905, PAR, NMC, T23, 5057/1905.

49. Analysis of Blue Book recordings of inyanga licenses from 1891 to 1906, complete record.

50. "Campbell Watt to SNA," September 16, 1910, PAR, NMC, T23, 1208/191.

51. "SAP: RE: Ghobozi to Natal Medical Council," November 23, 1914, PAR, NMC, T24.

52. Translation of Zulu article from *Izindaba Zabantu* dated March 15, 1915, and entitled "Doctors for Killing People," PAR, NMC, T24.

53. "Doctors' Correspondence," PAR, NMC, T12.

54. J. McCord, *Patients*, 68.

55. Ibid., 73.

56. Ibid., 250–52.

57. Evidence of a former African employee who worked there: "Rex v. Mafavuke Ngcobo," 174.

58. H. C. Lugg, *Life under a Zulu Shield* (Pietermaritzburg: Shuter and Shooter, 1975), 29.

59. *Natal's Blue Books* (1874–1910) record numerous district surgeons' complaints about African resistance to attending biomedical doctors. Only 3 percent of Africans in the rural areas of Natal and Zululand were said to visit biomedical doctors in 1928. "Report of Committee appointed to investigate and report on 'Health of Durban Natives' for Durban Joint Council of Europeans and Natives," September 1928, PAR, 3/DBN, 4/1/2/1219, 1.

60. *Blue Book of Native Affairs 1903*, report from Ndwedwe.

61. "Misc.," SAB, NTS, 6705, f 1/315.

62. J. McCord argues that many came to his dispensary instead of the government hospital because it did not have the same negative associations (*Patients*, 102, 103, 115). See *Supplement to the Blue Book for the Colony of Natal, Department Report 1886*, B26. Marks claims an increase in African use of hospitals followed the high African mortality rates in the 1930s and '40s (*Divided Sisterhood*, 11).

63. Matheson, "Social Medicine."

64. "Campbell Watt to SNA," September 16, 1910, PAR, NMC, T23, 1208/1911.

65. P. La Hausse, "The Struggle for the City: Alcohol, the Ematsheni, and Popular Culture in Durban, 1902–1936" (master's thesis, University of Cape Town, 1984), 120–22.

66. "To Town clerk from Provincial Secretary Maduna," August 31, 1925, PAR, 3/DBN, 4/1/2/1218, file 1.

67. Analysis of Blue Book records for numbers of licensed inyangas. Unfortunately the number of licenses were not recorded every year.

68. Jali: J. McCord, *Patients*, 285, 275.

69. Gordon to Natal Medical Council, December 15, 1924, PAR, NTS, 9301, 1/376, 2.

70. Callaway, *Religious System*; Bryant, *Zulu Medicine*.

71. A. W. Burton, "Medical Practitioners and Dispensing," *The Journal of the Medical Association of South Africa (JMASA)* 1, no. 19 (1927): 497–502.

72. W. G. Stafford, *Principles of Native Law* (Johannesburg: 1955), 205.

73. "CNC to SNA," Pretoria, May 9, 1931, PAR, NTS, 9301, 1/376, 3.

74. Circular no. 54 of the Natal Pharmaceutical Society, July 16, 1938, PAR, 1/DBN, 1/1/1/2/2.

75. Minutes of the Natal Pharmaceutical Society, 1908–23.

76. J. McCord, *Patients*, 268.

77. Shapiro, "Doctors or Medical Aides."

78. J. McCord, "Zulu Witch Doctor," 122.

79. Hay-Michel, "Native Medical Practitioners and European Patients," *JMASA* 1, no. 19 (1927).

80. Ibid.

81. Monamodi, "Medical Doctors," 285.

82. Ibid., 213.

83. J. A. du Toit, "Letter to the Editor: Native Doctors and European Patients," *JMASA* 2, no. 2 (1928): 45.

84. "Secretary for Public Health to Barrett," February 18, 1921, PAR, NTS, 9301, 1/376, 2.

85. H. A. Moffat, "Letter to the Editor: Native Doctors and European Patients," *JMASA* 1, no. 24 (1927): 659.

86. du Toit, "Letter."

87. Marks, *Divided Sisterhood*, 60.

88. P. Starr, *The Social Transformation of American Medicine* (New York: Basic Books, 1982).

89. "To NC, Nqutu from Bennett (solicitor)," June 2, 1937, SAB, GES, 1786, 25/30H.

90. "Secretary of Public Health to Minister of Public Health," May 4, 1933, PAR, NTS, 9301, 1/376, 1. (This does not include Zululand, which had more inyangas per person than Natal.)

91. I. Alexander mentions in his court case that there are only four people in Durban with inyanga licenses. (1940) PAR, 1/DBN, 1/1/1/2/2; 1931 numbers from miscellaneous paper in PAR, CNC, 50A, February 1931; population figures from P. Maylam and I. Edwards, *The Peoples' City* (Pietermaritzburg: University of Natal Press, 1996), 16.

92. Union of South Africa, *Natal Code of Native Law* (Pietermaritzburg, 1932), 24–25.

93. "Undersecretary for Native Affairs to CNC Natal," July 11, 1921, PAR, NTS, 9301, 1/376, 2. "Restrictions on Herbalists," *Natal Mercury*, August 2, 1939.

94. "CNC Natal to SNA Pretoria," September 30, 1938, PAR, GES, 1787, 25/30K; "Ngcobo to Minister of Native Affairs," March 2, 1931. PAR, NTS, 9301, 1/376, 3; "CNC Natal to SNA," May 9, 1931, PAR, NTS, 9301, 1/376, 3.

95. Stafford, *Principles of Native Law*, 205.

96. This committee formed in 1931, *The African Chemist and Druggist* 10, no. 107 (1931): 24. Unfortunately the minutes from 1923 to 1937 are missing from the Natal Pharmaceutical Society. The minutes from 1937 to 1945, however, show a very active Native and European Herbalist Committee.

97. Minutes of the Natal Pharmaceutical Society, held by NPS in Durban, Minute book of the Business Section 1935–1940. "NPS Circular 100, 22 December 1939."

98. Covered briefly in the executive minutes for March, April, and May of 1940. Minutes of the Natal Pharmaceutical Society, Minute Book—General and Executive 1937–1940.

99. "Ngcobo to Minister of Native Affairs," March 2, 1931, PAR, GES, 25/30E, 1784.

100. Kwa Ndlulamiti pamphlet PAR, NTS, 9303, 8/376; former president testified at Israel Alexander's trial in 1940, and see pamphlets in PAR, 1/DBN, 1/1/1/2/2.

101. Examples: KwaFleming Johnson and Seabanks Pharmacy from PAR, 1/DBN 1/1/1/2/2; KwaNdhlulamiti, SAB, NTS, 9303, 8/376; "Dr. Joffres Corrective Mixtures for Females," from J. Parle, "States of Mind: Mental Illness and the Quest for Mental Health in Natal and Zululand, 1868–1918" (PhD diss., University of KwaZulu-Natal, 2004), 276.

102. Testimony at I. Alexander trial, PAR, 1/DBN, 1/1/1/2/2.

103. Kwa Ndlulamiti pamphlet, PAR, NTS, 9303, 8/376.

104. Bengt Sundkler, who also acted as the Postal Agent of a mission district in northern Zululand, noted a steep increase in patent medicines in the early 1940s. Sundkler, *Bantu Prophets*, 222.

105. No author, "Natives Have Own 'Medical' Body," *Johannesburg Sunday Times*, April 10, 1938.

106. Ibid.; ""Native Medical Association"; Acutt and Worthington to Native Commissioner, Weenen, June 2, 1938, PAR, NTS, 9301, 1/376, 3; Alister to Post Commander, Weenen, May 11, 1938, PAR, NTS, 9301, 1/376, 3.

107. Quote from Cele to Minister of Native Affairs, February 12, 1931, PAR, NTS, 9301, 1/376, 3; Dube to CNC, October 7, 1938, PAR, NTS, 9301, 1/376, 3; SNA to Alport, December 22, 1941, PAR, NTS, 9301, 1/376, 3; "Restrictions on Herbalists," *Natal Mercury*, August 2, 1939.

108. For a fuller discussion of the African elite see S. Marks, *The Ambiguities of Dependence in South Africa* (Baltimore: Johns Hopkins University Press, 1986).

109. "Natives Have Own 'Medical' Body."

110. "Alister to Post Commander, South African Police," Weenen, May 11, 1938, PAR, NTS, 9301, 1/376, 3.

111. "Ngcobo to Minister of Native Affairs," Cape Town, March 2, 1931, PAR, NTS, 9301, 1/376, 3.

112. "Mazibuko and Ngcobo to CNC Natal," April 20, 1931, PAR, NTS, 9301, 1/376, 3.

113. "Natives Have Own 'Medical' Body."

114. Ibid.

115. In areas outside of Natal, the Dingaka Association (1928–circa 1960) based in the Transvaal organized male and female healers.

116. "Rex v. Mafavuke Ngcobo," SAB, GES, 1788, 25/30M, 318.

117. "Note" from CNC, CNC 43/25; "Dube to CNC," October 7, 1938, PAR, NTS, 9301, 1/376, 3.

118. "Medicine Man's Appeal Fails," *Daily News*, March 17, 1941. Also see "To Minister of Health from T. J. D'Alton," SAB, GES, 1785, August 29, 1936.

119. "Native Medical Association."

120. Vail, *Creation of Tribalism*, 15.

121. "Herbalist's £200 Daily Turnover," *Sunday Times*. Clipping not dated. Bantu newpapers clipping book 30, page 20, housed at the Killie Campbell Library, Durban.

CHAPTER 5: AFRICAN-INDIAN ENCOUNTERS

1. Please note that the term "Indian" has been problematized in the introduction and that I fully recognize the heterogeneity of this community.

2. "To Colonial Secretary from Parasoo Ramoodoo," dated March 18, 1905, and "To Colonial Secretary from Parasoo Ramoodoo," dated May 25, 1905. PAR, CSO, 1785, 1905/2373.

3. "Valuable missing after exorcism," *Sunday Times*, April 23, 2000; "Tokoloshe Blamed for Mom's Sex Torment," *Sunday Times*, July 29, 2001; the 2001 Carte Blanche documentary described at www.tokoloshe.tk (accessed June 27, 2004); "Strange Spirit Has Put Us All through Hell," *The Post*, March 17, 2004, www .thepost.co.za (accessed June 27, 2004). For *ufufunyane* possession see J. Parle, "Witchcraft or Madness? The *Amandiki* of Zululand, 1894–1914," *Journal of Southern African Studies* 29, no. 1 (March 2003): 105–32.

4. While there have been Indian and white inyangas since the nineteenth century, Indian and white isangomas seem to be a more recent and rare phenomenon.

5. A. G. Desai, "A Context for Violence: Social and Historical Underpinnings of Indo-African Violence in a South African Community" (PhD dissertation, Michigan State University, 1993). P. Jain, *Indians in South Africa* (Delhi: Kalinga, 1993). A. Sookdeo, "The Transformation of Ethnic Identities: The Case of 'Coloured' and Indian Africans," *The Journal of Ethnic Studies* 15, no. 4 (1988): 69–83.

6. During the riots 142 people died; 1,087 people were injured; one factory, 58 stores, and 247 dwellings were destroyed. B. Pachai and S. Bhana, *A Documentary History of Indian South Africans* (Cape Town: Hoover Institution Press, 1984), 208. For an analysis see P. Jain, *Indians in South Africa*, 47–51.

7. The Ghandi-Luthuli Documentation Centre in Durban houses a limited collection of documents related to KwaZulu-Natal history and also displays small exhibits that emphasize Indian and African unity. More recently, attempts have been made to create the Institute for Interethnic Studies, a center to foster better community relations between Indians and Africans. The re-established Phoenix Settlement published a pamphlet in English and Zulu that highlighted the works of Ghandhi and aimed to "help heal these fractures" between the Indian and African community. *The Opinion*, October 15, 2000.

8. UCLA Special Collections Department, Collection 1343, Hilda Kuper Papers (hereafter Kuper Papers), box 1, budget folder.

9. M. D. North-Coombes, "Indentured Labour in the Sugar Industries of Natal and Mauritius, 1834–1910," in *Essays on Indentured Indians in Natal*, ed. S. Bhana (Yorkshire: Peepal Tree Press, 1990), 62.

10. Ibid., 62–63.

11. Ibid., 67–71.

12. Quote by L. Govender: "We learned it [African therapeutics] when we were working together on the sugar plantations." S. Nesvag, "D'Urbanised Tradition: The Restructuring and Development of the *Muthi* Trade in Durban" (master's thesis, University of Natal, Durban, 1999), 95. Oral testimony about African and Indian women working side by side. "Indentured Service," *Natal Witness*, June 23, 2004.

13. While the literature on this skewed gender ratio has emphasized the pressures on a few Indian women to service many men domestically and sexually, it does not even consider miscegenation as a possibility. J. Beall, "Women under Indenture in Colonial Natal, 1860s–1911," in *South Asians Overseas: Migration and ethnicity*, ed. C. Clark, C. Peach, and S. Vertoure (Cambridge: Cambridge University Press, 1990), 65–69; H. Kuper, "'Strangers' in Plural Societies," in *Pluralism in Africa*, ed. L. Kuper and M. G. Smith (Berkeley and Los Angeles: University of California Press, 1969), 256–58. Kuper insists that cultural differences between Indians and Africans were too great for this to even be an option. This view seems much influenced by the time period in which she was soliciting Indian opinion. This is a difficult phenomenon to trace considering the lack of records; the cost of an £5 marriage license meant that even most Indian-Indian marriages were not state sanctioned. Considering R. Elphick and R. Shell's research on miscegenation in the early Cape and all the various factors involved, the possibility of Indian-African miscegenation seems quite reasonable. R. Elphick and R. Shell, "Intergroup Relations: Khoikhoi, Settlers, Slaves and Free blacks, 1652–1795," in *The Shaping of South African Society, 1652–1820*, ed. R. Elphick and H. Giliomee (Cape Town: Longman, 1979), 194–204.

14. Kuper, "'Strangers' in Plural Societies," 256.

15. *JSA*, 5:235–36, 246. Where Indians took over African land, such as Umzinto, however, Qalizwe is emphatic that such relations would never occur.

16. Beall, "Women under Indenture," 65.

17. The earliest mention of an Indian living under African rule comes from the 1850s—a decade before indenture—and clearly refers to a man from the Cape. Spohr, *Natal Diaries*, 68–70. For later references see PAR, SNA, I/1/141, 1891/513; PAR, SNA, I/1/49, 1561/1881; PAR, CNC, 112, 1913/390. To my knowledge no one has looked at the phenomenon of Indians living under African chiefs or how these relationships worked. Did Indians live as subjects to a chief as Africans did, or were they seen as "renters" with different social and political obligations?

18. "Report on matters and matters of interest that occurred in the above period," May 19–June 16, 1910, PAR, DPH, 36, miscellaneous.

19. See discussion of Togt labour recorded by James Stuart in 1902 with respondant Sisekelo. *JSA*, 5:360–64. Also an examination of Braby's directory for Durban and Pietermaritzburg shows Africans and Indians living within close proximity. *Braby's Natal Directory* (Braby Publishing). A good example of Indian-African encounters in Durban is Chittenden's description of the "Oriental" market, eating house, and muthi market on Grey Street. G. Chittenden, "Durban Delhi," *South African Railways and Harbours Magazine* (October 1915): 925–30. Also for a later period see 3/DBN, 4/1/2/1220, 11–12, "Durban Submission to Native Economic Commission 1930."

20. By 1900 they dominated the bulk of African trade. J. Brain, "Natal's Indians, 1860–1910," in *Natal and Zululand*, ed. Duminy and Guest, 259.

21. Kuper, "'Strangers' in Plural Societies," 258.

22. *Report of the Natal Native Commission, 1881–82*, 109.

23. Ibid., 384.

24. *JSA*, 5:360–63.

25. R. Mesthrie, "New Lights from Old Languages: Indian Languages and the Experience of Indentureship in South Africa," in *Essays on Indentured Indians in Natal*, ed. S. Bhana, 194–96.

26. *JSA*, 1:447–48, 157; *JSA*, 5:255–56.

27. Prashad, *Everybody Was Kungfu Fighting*, 94.

28. Dr. Xuma, who later became president of the ANC, stated in 1930: "The Indian cannot make common cause with the African without alienating the right of intervention on their behalf on the part of the Government of India." This sentiment reflected that of white South Africans who viewed Indians as temporary rather than permanent residents of South Africa. As quoted in A. Sookdeo, "Transformation of Ethnic Identities: The Case of 'Coloured' and Indian Africans, *Journal of Ethnic Studies* 15, no. 4 (1988): 72.

29. *The Opinion*, October 15, 2000.

30. All this information comes from Sookdeo, "Transformation," 75–77.

31. Prashad, *Everybody Was Kungfu Fighting*, 71.

32. Ibid., 70–96.

33. R. Jain, *Indian Communities Abroad* (New Delhi: Monahar, 1993), 12–13.

34. Ebr.-Vally, *Kala Pani*, 132–38; R. Jain, *Indian Communities Abroad*, 4.

35. Ebr.-Vally, *Kala Pani*, 133–34.

36. The loss of caste in certain diaporic communities may be similar to the process that led lower or unscheduled caste groups to convert from Hinduism to Islam and Christiantiy.

37. Prashad, *Everybody Was Kungfu Fighting*, 94.

38. For instance upper-class Indians exchanged trade favors such as the the 1925 Class Areas bill, which promoted segregation, and the Natal branch of the South African Indian Congress agreed to "repatriate" Indians of the poorer classes. Sookdeo, "Transformation," 72–73. Likewise the African community felt betrayed by African elites like D. T. Jabavu, who apparently agreed to the Natives Land Tenure and Representation Bill, which would retain a qualified franchise for elite Christian Africans. Sookdeo, "Transformation," 75.

39. Kuper, "'Strangers' in Plural Societies," 257–58; P. van den Berghe, "Asians in East and South Africa," in *Race and Ethnicity* (New York: Basic Books, 1970); R. Morrell, J. Wright, and S. Meintjes, "Colonialism and the Establishment of White Domination, 1840–1890," in *Political Economy and Identities in KwaZulu-Natal*, ed. R. Morrell (Durban: Indicator Press, 1996), 41.

40. In a study of Indian psychiatric patients in 1986, approximately 50 percent said that they had visited an Indian healer. K. Bhana, "Indian Indigenous Healers," *South African Medical Journal* 70, no. 4 (1986): 221–23.

41. Dr. V. Naiker, founder of the recent Ayurvedic Society of South Africa, interview, June 24, 2002, Durban. Nowadays many identify Ayurveda as the "rational," nonsecular practice of Indian medicine divorced from home remedies and other more spiritual aspects of Indian medicine. Jean Langford challenges this notion arguing that this division is a more recent invention and not one found in nineteenth-century Ayurveda. Jean Langford, *Fluent Bodies: Ayurvedic Remedies for Postcolonial Imbalance* (Durham, NC: Duke University Press, 2002).

42. Examples too numerous to note are found throughout the Kuper Papers. Ravi Govender, informal interview, June 11, 2002, Durban.

43. Z. Mayat, ed., *Indian Delights* (Durban: Women's Cultural Group, 1998; first ed., 1961), 9, 299–305. I thank Shobana Shankar for the suggestion of looking at Indian cookbooks.

44. Kuper Papers, box 1, "Budget Folder," results of budget survey.

45. Ibid., 236–61.

46. Ibid., 217–35.

47. T. Pillay, interview, June 20, 2002, Ladysmith; F. Soofie, great-great-grandson of Soofie Saheb, interview, June 28, 2002, Durban.

48. T. Pillay interview; S. Pillay, interview, June 20, 2002, Newcastle.

49. This comes from an economic survey of the Indian community conducted in the 1950s, as reported in H. Kuper and F. Meer, "Indian Elites in Natal, South Africa," in *University of Natal Institute for Social Research Proceedings* (July 1956): 136.

50. 1881 *Natal Native Commission*, 88, 384. One mentions "curry," the other "medicine." Quote from "Report by Medical Officer, Durban, September 3, 1881," 1881 *Blue Book on Native Affairs*, 72.

51. 1892, Insinga: PAR, SNA, 1/1/64; 1893, Umgeni: PAR, SNA, I/1/168, 354/93.

52. "Minutes of the Natal Medical Council," February 11, 1898, PAR, NMC, T1.

53. "Translation of report of Krantzkrop," no. 1, November 26, 1908, PAR, 1/KRK, 3/1/9.

54. L. Govender interview.

55. L. Naidoo, interview, April 1, 1998, Pietermaritzburg.

56. Ayurveda is a form of medicine which has developed over thousands of years on the Indian subcontinent, it includes not only herbal and mineral remedies, but message and bleeding, and using food to bring the body back into balance.

57. T. Pillay interview.

58. A. Ngwanye, informal interview, June 24, 2002, Durban; A. Naiker interview, June 26, 2002, Durban.

59. A. Smith, *A Contribution to South African Materia Medica Chiefly from Plants in Use among the Natives* (Cape Town: Juta, 1895, 3rd ed.), 227.

60. PAR, 1/KRK, 3/1/9. He charged ten shillings for treatment, which seemed to be the going rate for this area.

61. "Remarkable Quackery," *The Probe* 4, no. 8 (June 1, 1914): 153.

62. L. Naidoo interview, August 20, 1998, Howick Falls.

63. Christina Sibiya mentions that King Solomon seeks out three African inyangas and one Durban Indian inyanga to discover who had bewitched Christina's son. The time period referenced in 1927. R. H. Reyher, *Zulu Woman*, 147–48. Oral histories also point to other Indians involved in the *muthi* business as offering such diagnoses.

64. L. Govender interview.

65. This changed in 1928 with the passage of the Medical, Dental, Pharmaceutical Act which transferred the right of licencing to the department of public health.

66. "Application by free Indian Seetal," PAR, CNC, 112, 1913/390.

67. Chinnavadu: PAR, CNC, 284, 1917/1992/93; Sevuthean: PAR, CNC, 324, 1918/1451.

68. Archival sources point to muthi markets from 1906 to 1920; oral sources, however, claim that muthi, among other items, had been sold at the Mona market since the time of the Zulu kingdom. Nesvag, "D'Urbanised Tradition," 61; *1906–07 Native Commission*, Migidhlana of Eshowe, 895.

69. Chittenden, "Durban Delhi," 928.

70. For a historical account of street trading see G. Vahed, "Control and Repression: The Plight of Indian Hawkers and Flower Sellers in the Durban CBD, 1910–1948," *International Journal of African Historical Studies* 32, no. 1 (1999): 19–48; S. Nesvag, "Street Trading from Apartheid to Post-Apartheid: More Birds in the Cornfield?" *International Journal of Sociology and Social Policy* 21, nos. 3/4 (2000): 34–63.

71. *Braby's Durban Directory 1930* (Braby's Publishing), listed under Queen Street.

72. F. G. Cawston, "Native Medicines in Natal," *South African Medical Journal* 7, no. 11 (1933): 371.

73. L. Naidoo interview, April 1, 1998.

74. For a discussion of "pollution," particularly among female traders, see Nesvag, "D'Urbanised Tradition," 170–72.

75. L. Govender interview and *Braby's Durban Directory* 1936, 1937. C. V. Pillay is actually listed in Braby's Directory in 1937 and for several years after as the proprietor of the Victoria property, perhaps indicating that he had signed and was responsible for the lease.

76. According to P. Cele, Durban administrator, "Mr. Chester explained that no one must display their herbs outdoors; they must practice indoors." P. Cele interview. Also *Braby Durban Directory* lists Cele's shop on Leopold Street starting in 1939.

77. "To Minister of Health from T. J. D'Alton," SAB, GES, 1785.

78. *Braby's Durban Directory* 1936, 1945.

79. P. Cele interview.

80. A. Naiken, interview.

81. Mpongose, interview, October 30, 1998, Durban. In her interview (December 23, 1998, Durban), Victoria Street Market Woman says similiarly, "Indians? They use the same medicines we use . . . Some of the Indians that was staying with us at the farms learned those, and now they doctors, they *inyangas* also. Not *sangomas*, though some of them are *sangomas*."

82. As quoted in Nesvag, "D'Urbanised Tradition," 95.

83. P. Cele interview. Also see Mpongose interview and S. Shange interview, 1979, conducted and housed at Oral History at KCL.

84. Nesvag, "D'Urbanised Tradition," 96.

85. A. Naiken interview.

86. "Rex v. Mafavuke Ngcobo" (1940), SAB, GES, 1788, 25/30M, 240–320.

87. A. Naiken interview. Also supported by T. Pillay interview.

88. *JSA*, 3:182, evidence of Mkando.

89. J. Colenso, *Zulu-English Dictionary*, 254.

90. Naidoo interview, August 28, 1998, Pietermaritzburg; T. Pillay interview; Naiken interview.

91. T. Pillay interview; S. Pillay interview.

92. Holy basil grows wild throughout Natal and is not sold in muthi shops. In L. Naidoo's shop he sold Indian herbs for facial creams and love potions. L. Naidoo interview, August 28, 1998, Pietermaritzburg. These are the same things that Mpongose claims are missing from local therapeutics. Mpongose interview.

93. Literally "Indian aloe." L. Naidoo interview, August 28, 1998. Mpongose interview.

94. J. Mbuyazi interview. Indian female inyangas are in fact a more recent phenomenon and obviously did not factor into this man's thinking.

95. F. Soofie interview. He showed me photographs from 1951 of this festival and confirmed as many others in this community had that Africans had been

coming to the Badsha Peer shrine from the early twentieth century. Soofie attributed this to the fact that Sufis don't charge exhorbitant fees. Also the museum at the Riverside Soofie Saheb Shrine has a number of pictures from the first half of the twentieth century that show Africans at various Sufi events, though they are few in number.

96. Notebook dated 1953 referring to an event that took place in five to six years ago. Kuper Papers, 1343, box 15, notebook VIIx.

97. Notebook dated March 14, 1954, Kuper Papers, 1343, box 14, notebook 12AA; Kuper Papers, 1343, box 15, notebook VIIx. Indians seeing Africans healers: Kuper Papers; also see example that shows these same phenomenon were still occuring in the early 1970s. Seedat, "The Zanzibaris in Durban."

98. Kuper Papers, box 12, notebook 5D, 1954, *takathi* seems to refer to an "Indian form of witchcraft — 'to tie your spirit.'" Wohja-Kuper Papers, box 15, notebook VIIx.

99. 1953 Kuper Papers, box 15, notebook 3AA.

100. Shange interview.

101. R. Pillay, interview, August 19, 1953, Kuper Papers, Box 9. Interestingly, the notebook transcript is somewhat different than what ends up in Kuper's book. In H. Kuper, *Indian People in Natal* (Pietermartizburg: Natal University Press, 1960), the child is nine years old, the first doctor is a biomedical doctor rather than an Indian diviner, and fewer African healers are mentioned.

102. Kuper, *Indian People in Natal*, 258–59.

103. Ngubane, *Body and Mind*, 144–46.

104. In the early 1970s Zanzibaris were reputed to be the richest members of their communities, and were said to serve both African and Indian communities. Seedat, "The Zanzibaris in Durban," 99. Surveys were conducted in 1971.

105. Applications for General Dealers Licenses increased with the restriction of *inyanga* licenses as prescribed in the 1928 Medical, Dental, and Pharmacy Act. SAB,GES, 1785, "To Minister of Health from T. J. D'Alton," August 29, 1936.

106. "Note" from CNC discusses need to "preserve the genuine native *nyanga*," PAR, CNC, 43/25; "Dube to CNC" presents resolutions of 1938 Natal Native Congress, which includes rights of Africans to sell to each other without "foreign races standing in the way," October 7, 1938, CAD, NTS, 9301, 1/376,.

107. PAR, NTS, 2/376, part 2, September 1948.

108. Indian herbal remedies such as honey, caster oil, cloves, ginger, garlic, mother's milk, betel leaf, areca nut, turmeric, and syringa leaves were used in numerous therapies as well as health rituals surrounding birth, menstruation, and marriage. Examples to numerous to note are found throughout Kuper Papers; R. Govender interview.

109. *Hazrath Soofie Saheb (Rahmatullah Alai) and his Khanqahs* (Durban: n.d., circa 1998), 70. Dawakhana clinic, 83. This book was printed by and for the Soofie Saheb community.

110. Cawston, "Native Medicines in Natal," 370. These herbs included *chavica betle* and *areca oberacea*, and *areca catechu linn*. "For sweetening the breath and

for preserving the gums, Indians sometimes use a resin prepared from the bark of *pistachia lintiscus,* or the order of *anacardiacea,* which grows in Natal."

111. "Minutes of the Natal Pharmaceutical Society, 6 April 1937," housed in the Natal Pharmaceutical Society, Durban, Minute Book of the Business Section.

112. Kuper Papers, box 9 (book 2aa, 77). Kuper Papers, "Survey of Traders in the Township of Tongaat," box 11.

113. A. Naiken interview.

114. K. Flint, "Competition, Race, and Professionalization."

115. L. Govender interview; A. Naiken interview; L. Naidoo interview, July 29, 1998.

116. This is my informal observation after interviewing over thirty healers in the area of KwaZulu-Natal in 1997–98.

EPILOGUE

1. M. Last and P. Chavunduka, eds., *The Professionalisation of African Medicine* (Manchester: Manchester University Press, 1986); D. Simmons, "Of Markets and Medicines," in *Borders and Healers,* ed. T. Luedke and H. West (Bloomington: Indiana University Press, 2006), 65–81; J. Stephan, "Traditional and Alternative Systems of Medicine" in *International Digest of Health Legislation* 36, no. 2 (1985); C. Leslie, "Medical Pluralism in World Perspective," in *Social Science and Medicine* 14B (1980), 191–95.

2. A series of articles address the AIDS epidemic specifically in Southern Africa, many of which cover new collaborative efforts between traditional healers and biomedicine.

3. K. Pretorius, "Institute for Traditional Medicines," April 8, 2004, http://www.southafrica.info/ess_info/sa_glance/health/traditionalmedicine.htm (last accessed December 15, 2006).

4. M. Colvin et al, "Integrating traditional healers into a tuberculosis control programme in Hlabisa, South Africa," *MRC Policy Brief* no. 5, December 2001.

5. South Africa's National Research Foundation began funding on IKS in the year 2000. G. Augusto, "Knowing Differently, Innovating Together? An Exploratory Case Study of Trans-Epistemic Interaction in a South African Bioprospecting Program" (PhD diss., George Washington University, 2004), 144.

6. CSIR in fact boasts of hoodia on its bioprospecting website: http://www.csir.co.za/plsql/ptl0002/PTL0002_PGE057_RESEARCH?DIVISION_NO=1010012&PROGRAM_NO=3410026 (accessed February 15, 2007).

7. M. Sayagnes, "South Africa: Indigenous Group Wins Rights to Its Healing Herbs." www.corpwatch.org/article (accessed September 29, 2005).

8. V. Shiva, *Biopiracy.*

9. My emphasis, as quoted in D. Posey and G. Dutfield, *Beyond Intellectual Property* (Ottawa: International Development and Research Centre, 1996), 103.

10. A. Barnett, "Fat Windfall for the San," *South African Mail and Guardian*, April 6, 2002; "Bushmen to Win Royalties from Slimming Drug," *South African Mail and Guardian*, March 27, 2003; "San Stand to Make Millions off Fat of the Land," *South African Mail and Guardian*, July 31, 2003; "Protesting Traditional Knowledge: The San and Hoodia," *Bulletin of the World Health Organization* 84, no. 5 (2006).

11. M. Duenwald, "An Appetite Killer for a Killer Appetite? Not Yet," *The New York Times*, April 19, 2005. L. Johannes, "Hoodia's Hunger Claims," *The Wall Street Journal*, December 13, 2005.

12. "Unilever Gains Exclusive Rights to Phytopharm's Hoodia Extract," *Breaking News on Supplements and Nutrition in Europe*, December 15, 2004; See 60 *Minutes* report at: http://www.cbsnews.com/stories/2004/11/18/60minutes/main656458 .shtml (last accessed December 15, 2006).

13. To date there has been only one unpublished human trial on hoodia conducted by Phytopharm itself. Given that the study included only twelve subjects over a short three-month period and has not been subjected to peer review, the scientific community has remained rather skeptical. Likewise, tests of ten supplements claiming to contain hoodia showed that only two brands contained the active compound P57 said to act as an appetite suppressant. B. Avula et al., "Determination of the Appetite Suppressant P57 in Hoodia Gordonii Plant Extracts and Dietary Supplements by Liquid Chromatorgraphy/Electrospray Ionization Mass Spectrometry (LC-MSD-TOF) and LC-UV Methods," *Journal of AOAC International* 89, no. 3 (2006).

14. Given that hoodia is a slow-growing succulent that takes decades to mature and is difficult to grow outside its native habitat of the Kalahari Desert, it must be carefully harvested to maintain sustainability. Avula et al., "Appetite Suppressant," 606–11.

15. A farmer was recently arrested for illegally harvesting two tons of the hoodia plant, said to be worth two million rand. S. Kwon Hoo, "Farmer Faces R6million Fine after Huge Hoodia Haul," *Diamond Fields Advertiser*, May 25, 2006.

16. A good example is J. McCord, " Zulu Witch Doctor"; a more recent example of this in east Africa is S. Langwick, "Ethnographies of Medicine," in *Borders and Healers*.

17. Convention on Biological Diversity site: www.biodiv.org/worl/map.asp?ctr =za (accessed February 15, 2007).

18. This includes initiatives by the World Health Organization, which declared the need to recognize and utilize indigenous medicines as early as 1977, and the more recent 1992 Convention on Biodiversity. Lawyers for the San recently appealed to the Convention on Biodiversity to prevent Germany and Switzerland from selling unauthorized hoodia. WHO, "Protesting Traditional Knowledge."

19. Posey and Dutfield, *Beyond Intellectual Property*, 50–57.

20. Statistics from South African Department of Health. This is a compilation of statistics collected by the Actuarial Society of South African, Centre for Actuarial

Research and the Medical Research Council, posted at: http://www.doh.gov.za/docs/reports/2006/summary.html (last accessed March 3, 2007). These are the same statistics used by UNAIDS to determine the rates of the disease in South Africa.

21. There have been a number of good sources on the historical contexts of the HIV/AIDS debate that cover these factors in greater detail. An entire journal dedicated to this subject in South Africa is *African Studies* 1 (2002); see also C. Burns, "Sex Lessons from the Past?" *Agenda* 29 (1996): 79–91; M. Mahoney and J. Parle, "An Ambiguous Sexual Revolution? Sexual Change and Intra-generational Conflict in Colonial Natal," *South African Historical Journal* no. 50 (2004): 134–51; M. Mbali, "A Medical History 'From Below': A Critical Review of New Literature on Changes in African Culture in South Africa and STD and AIDS EPI," *Journal of Natal and Zulu History* 21 (2003): 77–93; A. Butler, "South Africa's HIV/Aids Policy, 1994–2004: How Can It Be Explained?" *African Affairs* 104, no. 417 (2005): 591–614. Also see www.avert.org/aidssouthafrica.htm (last accessed March 3, 2007), and J. Iliffe, *The African AIDS Epidemic* (Athens: Ohio University Press, 2006).

22. Wellings et al., "Sexual Behaviour in Context: A Global Perspective," *The Lancet*, November 11, 2006, 368, 9548, 1706–28.

23. For information on this apartheid program see H. Bradford, "'Her Body, Her Life': 150 Years of Abortion in South Africa," unpublished paper, 1991; B. Klugman, "The Politics of Contraception in South Africa," *Women's Studies International Forum* 13, no. 3 (1990): 261–71; H. Rees, "The Abortion Debate in South Africa," *Critical Health* 34 (June 1991): 20–26.

24. K. Jochelson, "Sexually Transmitted Diseases in Nineteenth- and Twentieth-Century South Africa," in *Histories of Sexually Transmitted Diseases and HIV/AIDS in Sub-Saharan Africa*, ed. P. Setel et al. (Westport, CT: Greenwood, 1999), 217–44.

25. The long-lasting consequences of such actions could be seen as late as November 2004, when *The Washington Post* reported that although Anglo-America, a mining company, had offered free antiretroviral drugs for two years to all employees who reached a critical level of the disease, few had taken advantage of the offer. It reported that 75 percent of those in need had refused the drugs. Likewise a doctor at Johannesburg General Hospital estimated that one out of three patients who were offered free antiretroviral therapies refused them. Doctors have speculated that those rejecting such biomedical treatments did so out of fear of stigma, the noxious side effects of antiretrovirals, and scepticism of their efficacy. Admittedly the national program to distribute free antiretroviral therapies is fairly new, and while there has been scepticism, demand for such drugs has outstripped supplies. See C. Timberg, "S. Africans Shun a Remedy for AIDS," *Washington Post*, October 21, 2004.

26. Butler, "South Africa's HIV/AIDS Policy," 591–614.

27. Timberg, "S. Africans Shun."

28. Many of the healers I interviewed in KwaZulu-Natal in 1998, in both Durban and the rural area of Hlabisa, had recently participated in such workshops,

though it was clear that those in higher positions in the healing associations had had a greater likelihood of attending such a workshop. These workshops seem to have been fairly pervasive throughout this area and met with good success. Homsey et al., "Defining Minimum Standards of Practice for Incorporating African Traditional Medicine into HIV/AIDS Prevention, Care, and Support: A Region Initiative in Eastern and Southern Africa," *The Journal of Alternative and Complementary Medicine* 10, no. 5 (2004): 906. For knowledge base of healers regarding recent HIV knowledge of healers, see K. Peltzer et al., "HIV/AIDS/STI/TB Knowledge, Beliefs and Practices of Traditional Healers in KwaZulu-Natal, South Africa," *AIDS Care* 18, no. 6 (August 2006): 608–13.

29. South African Medical Research Council Policy Brief, no. 5, December 2001.

30. See SANAC homepage at http://www.info.gov.za/issues/hiv/sanac.htm (last accessed March 6, 2007).

31. http://www.doh.gov.za/docs/sp/2003/sp1022.html (last accessed March 4, 2007).

32. M. Phillips, "Persuading Africans to Take Their Herbs with Some Antivirals," *Wall Street Journal*, May 5, 2006.

33. M. Richter, "Discussion Paper Prepared for the Treatment Action Campaign and AIDS Law Project," November 2003, http://www.tac.org.za/Documents/ ResearchPapers/Traditional_Medicine_briefing.pdf (last accessed March 4, 2007).

34. Treatment Action Campaign's newsletter, *EqualTreatment*, http://www.tac .org.za/ET/EqualTreatmentMay2005Issue15.pdf. (last accessed March 4, 2007).

35. E. Mills et al., "Impact of African Herbal Medicines on Antiretroviral Metabolism," *AIDS* 19, no. 1 (2005): 95–97.

36. Some success in explaining these issues can be seen in D. McNeil, "AIDS Crisis Leaves Africa's Oldest Ways at a Loss," *The New York Times*, November 27, 2001.

37. C. Burns raises this same point for the earlier period of fertility control in South Africa in "Sex Lessons."

38. A. Cunningham, "The Herbal Medicine Trade," in *South Africa's Informal Economy*, ed. E. Preston-Whyte and C. Rogerson (Cape Town: Oxford University Press, 1991), 201–2.

39. Visitors to the contemporary herbal markets of Durban or Johannesburg will note that many of the sellers/collectors are in fact women. It is unclear when this profession became feminized. Ngcobo's court case mentions their presence by 1940, but pictures and visual descriptions of Durban's eMatcheni market from the 1930s do not mention or show women sellers. See "Rex v. Mafavuke Ngcobo," SAB, GES, 1788, 25/30M.

40. In 2002 Myles Manders told a *Los Angeles Times* reporter that the muthi trade is a $200-million-a-year business. He described the depletion of the pepper bark tree and wild ginger, which now must be collected in places like Zimbabwe and Mozambique. D. Maharaj, "The World: South African Healers Burgeoning," *Los Angeles Times*, October 27, 2002.

41. N. Crouch. "Sustaining livelihoods-annesting the erosion of ethnobotanical knowledge and related plant resources." *SANBI Biodiversity Series* 1 (2006): 53-56.

42. South African National Botanical Institute (SANBI) Biodiversity Series 1 (2006): 54.

43. Mthemjwa interivew, October 1998, Durban.

44. Devenish, "Negotiating Healing," 85.

45. Interviews with V. Mkize, 1998, 2002, Durban; and Nesvag, "D'Urbanised Tradition."

46. Q. Ntuli interview, June 26, 2002, KwaMuhle Museum, Durban.

47. Q. Ntuli interview, December 10, 1998, her Durban shop.

48. This information on the workings of the hospital is from S. S. Campbell, *Called to Heal: Traditional Healing Meets Modern Medicine in Southern Africa Today* (Johannesburg: Zebra, 1998): 98–105.

Glossary

NOTE ON GLOSSARY AND ORTHOGRAPHY

Given the colonial experience and the prominence of the English language in South Africa, but particularly in KwaZulu-Natal, a number of Zulu terms have been anglicized over the years and gained popular usage throughout South Africa. The Zulu word for medicine, *umuThi*, is thus popularly called "muthi" and sometimes spelled in a non-Zulu way as "muti." In order to avoid confusing those who are not familiar with the Zulu language I have taken the liberty of using these anglicized terms and anglicizing Zulu plurals by adding an *s*; thus many medicines would be "muthis." In quotations I have left the spelling of Zulu plurals as they appear in the original. In order to help the reader I have included a glossary below of Zulu terms with both the singular and plural forms. The spelling of terms here is in contemporary Zulu, though quotations from primary sources often reflect older orthographic traditions of Zulu spelling.

ZULU TERMS

Singular or Verb	Plural	Meaning
Assegai		spear
iBandla		council of elder statespersons
imBongi		bard or one who recites the history of the nation or praise poems for the chief or king

Singular or Verb	Plural	Meaning
izibongo		praises and history of the nation
ukuBula		to divine by smiting with sticks in response to an *isaNgoma*'s questions
uButho	*amaButho*	*sing:* soldier; *pl:* soldiers or regiment
iDlozi	*amaDlozi*	ancestral spirit
inDuna	*izinDuna*	a councilor or headman of a chief
	iziDwedwe	material containing body dirt which carries the essence of a person and is thus used for ritualistic purposes or is mixed with *muthi* for evil purposes
ukuGcaba		to make incisions in the skin and rub in medicine
isiGodlo	*iziGodlo*	private area of the king's enclosure that housed the king's wives, children, and concubines as well as the nation's *inKatha*
uGxa		initial fee of a doctor
umHlahlo	*imiHlahlo*	an assembly at which *isaNgomas* are asked to divine
uHlaka	*abaHlaka*	apprentice of a healer, who carries the *umuThi*
inKatha	*izinKatha*	ceremonial grass coil used to ensure the solidarity and strength of the chiefdom or kingdom
umKhando	*imiKhando*	finely ground Indian powders used in *umuThi*
uKholwa	*amaKholwa*	African Christian converts, usually fairly educated and better off economically

Singular or Verb	Plural	Meaning
umKhosi		First Fruits ceremony marked the eating of the season's first crops by the king, as well as the ritual empowerment of the king, nation, and soldiers by the king's doctors
umKhuhlane	*imiKhuhlane*	illness caused by natural origins
ukuKonza		to pledge allegiance to
iLobolo		gift exchange given to the maternal family at time of marriage
umLozi	*imiLozi*	whistling spirit which usually emanates from the roof of a structure
umlozikazana	*abalozikazana*	person who channels an *umLozi* or whistling spirit
iMpaka	*iziMpaka*	cat that acts as the familiar of an *umThakathi* (defined below)
isaNgoma/isaNuse	*izaNgoma/izaNuse*	healer who uses the power of the ancestors to divine or "smell out" the cause of illness and can intervene between the living and the *amaDlozi*
iNyanga	*iziNyanga*	specialist or healer who uses *umuthi* alone
inSila	*izinSila*	body dirt
umSutu	*abaSutu*	doctor who strengthens the army and nation
inTelezi	*izinTelezi*	protective medicine
umThakathi	*abaThakathi*	witch, evil-doer, or criminal; sometimes spelled *umthagathi*
umThandazi	*abaThandazi*	healers who use the power of the holy ghost and Christian prayer to heal

umuThi	*imiThi*	medicine; can include herbs, animals, minerals, and/or chemicals
isiThundu		vessel for carrying magical medicines, owned by powerful chiefs and kings
umTwasa	*abaTwasa*	an *iSangoma* initiate who is alerted to his/her calling by ailments particular to an *umTwasa*
ukuTwasa		to undergo illness/initiation to become an *iSangoma*

Bibliography

ARCHIVAL PRIMARY SOURCES

Pietermaritzburg Archive Repository (PAR), Pietermaritzburg, South Africa

Attorney General
Chief Native Commissioner
Colonial Secretary
Magistrate's and Commissioner's Reports:
    Bulwer
    Camperdown
    Empangeni
    Eshowe
    Howick
    Ixopo
    Kranskop
    Ladysmith
    Melmoth
    Nongoma
    Pietermaritzburg
    Weenen
Natal Medical Council, 1896–1930
Native High Court
Secretary of Native Affairs

Durban Archive Repository (DAR), Durban, South Africa

Bantu Administration Committee 1929–36
Bantu Advisory Board 1930–42

Durban Town Council
Native Advisory Board 1929–44
Native Affairs Department 1916–45
Native Affairs of the Borough 1915–33
Native Affairs of the City 1935–46
Public Health 1914–50

National Archive Repository (SAB), Pretoria, South Africa

Department of Health
Governor General
Native Affairs

Hilda Kuper Papers, UCLA Special Collections Department, Collection 1343

GOVERNMENT PUBLICATIONS

1931 *Durban Joint Council of Report of Committee Appointed to Investigate and Report of Health of Durban Natives.*
Blue Book on Native Affairs (1879–86, 1894–98, 1901–9).
Commission to inquire into the past and present state of the Kafirs. Proceedings and report of the Commission appointed to inquire into the past and present state of the Kafirs in the district of Natal, and to report upon their future government and to suggest such arrangements as will tend to secure the peace and welfare of the district: for the information of His Honour Lieutenant-Governor Pine: 1852–53.
Economic and Wage Commission 1925.
Evidence taken before the Natal Native Commission, 1881.
Natal Code of Native Law.
Natal Native Affairs Commission 1906–7.
Report of Native Churches Commission of 1925.
Report of the Coolie Commission 1872.
Report of the Commission of Leprosy.
Report of the Indian Immigrants Commission.
Report of the Natal Native Commission, 1881–82.
Report of the National European-Bantu Conference, Cape Town, February 6–9, 1929.
Report on Native Affairs in Natal, 1852–53.
South African Native Affairs Commission 1903–5 (volume 3 for Natal).
Select Committee on the Contagious Diseases Prevention Bill no. 19, 1890.

MEDICAL JOURNALS

The African Chemist and Druggist (1921–35).
The Journal of the Medical Association of South Africa (1927–31).

The Leech (1932–50).
The Probe (1910–14).
The South African Journal of Medical Sciences (1935–41).
The South African Medical Journal (1884–99, 1932–49).
The South African Medical Record (1903–26).
The South African Pharmaceutical Journal (1934–40).

NEWSPAPERS

Ilanga lase Natal
Inkanyiso yase Natal
Izindaba zabantu
Natal Mercury
Natal Witness

PUBLISHED SOURCES

Altick, R. *The Shows of London*. Cambridge, MA: Harvard University Press, 1978.
Anderson, B. *Imagined Communities: Reflections on the Origin and Spread of Nationalism*. London: Verso Editions, 1983.
Anonymous. "Medica Chiefly from Plants in Use Among the Natives by Andrew Smith." *South African Medical Journal* 4, no. 1 (1896).
Arnold, D. *Colonizing the Body: State Medicine and Epidemic Disease in Nineteenth-Century India*. Berkeley: University of California Press, 1993.
———, ed. *Imperial Medicine and Indigenous Societies*. Manchester: Manchester University Press, 1988.
Ashforth, A. *Madumo: A Man Bewitched*. Chicago: University of Chicago Press, 2005.
———. *Witchcraft, Violence, and Democracy in South Africa*. Chicago: University of Chicago Press, 2005.
Baldwin-Ragaven, L. *An Ambulance of the Wrong Color—Health Professionals, Human Rights and Ethics in South Africa*. Cape Town: University of Cape Town Press, 1999.
Barker, A. *Physic and Protocol among the Zulus*. Johannesburg: Institute for the Study of Man in Africa, 1972.
Barker, L. *A Year's Housekeeping in South Africa*. London: Macmillan, 1883. First published 1877.
Barter, C. *Alone among the Zulus: The Narrative of a Journey through the Zulu Country, South Africa*. Pietermaritzburg: University of Natal Press, 1995. First published 1866.
Baumann, G. *The Multicultural Riddle: Rethinking National, Ethnic, and Religious Identities*. NewYork: Routledge, 1999.
Bell, H. *Frontiers of Medicine in the Anglo-Egyptian Sudan*. Oxford: Clarendon Press, 1999.

Berger, I. "Rebels or Status-Seekers? Women as Spirit Mediums in East Africa." In *Women in Africa: Studies in Social and Economic Change*, edited by N. Hafkin and E. Bay, 157–81. Stanford: Stanford University Press, 1976.

Berglund, A. I. *Zulu Thought Patterns and Symbolism*. New York: Holmes and Meier, 1976.

Bhabha, H. *The Location of Culture*. London: Routledge, 1994.

Bhana, K. "Indian Indigenous Healers." *South African Medical Journal* 70, no. 4 (1986): 221–23.

Bhana, S., ed. *Essays on Indentured Indians in Natal*. Yorkshire: Peepal Tree Press, 1990.

Bird, J. *The Annals of Natal, 1495 to 1845*. 2 vols. Cape Town: C. Struik, 1965.

Blair, B. *Surgical Experiences in the Zulu and Transvaal Wars, 1879–1881*. Edinburgh, 1883.

Booth, A., ed. *Journal of Rev. George Champion, American Missionary in Zululand, 1835–9*. Cape Town: C. Struik, 1967.

Bourquin, S., ed. *Paulina Dlamini: Servant of Two Kings*. Pietermaritzburg: University of Natal Press, 1986.

Braby's Natal Directory.

Brain, J. "Health and Disease in White Settlers in Colonial Natal." *Natalia* 15 (1985).

Breman, J., ed. *Imperial Monkey Business: Racial Supremacy in Social Darwinist Theory and Colonial Practice*. Amsterdam: VU University Press, 1990.

Brookes, E. *White Rule in South Africa, 1830–1910: Varieties in Governmental Policies Affecting Africans*. Pietermaritzburg: University of Natal Press, 1974.

Bryant, A. T. *Olden Times in Zululand and Natal*. London: Longmans, Green; New York: Green, 1929.

———. "The Zulu Cult of the Dead." *Man* 17, no. 95 (1917): 140–45.

———. *A Zulu-English dictionary with notes on pronunciation, a revised orthography and derivations and cognate words from many languages; including also a vocabulary of Hlonipa words, tribal-names, etc., a synopsis of Zulu grammar and a concise history of the Zulu people from the most ancient times*. Pinetown: RSA, 1905.

———. *Zulu Medicine and Medicine Men*. Cape Town: C. Struit, 1966. First published 1909.

———. *The Zulu People as They Were before the White Man Came*. Pietermaritzburg: Shuter and Shooter, 1949.

Burke, T. *Lifebuoy Men, Lux Women*. Durham, NC: Duke University Press, 1996.

Burns, C. "Louisa Mvemve: A Woman's Advice to the Public on the Cure of Various Diseases." *Kronos: Journal of Cape History* 23 (1996): 108–34.

———. "'A Man Is a Clumsy Thing Who Does Not Know How to Handle a Sick Person': Aspects of the History of Masculinity and Race in the Shaping of Male Nursing in South Africa, 1900–1950." *Journal of Southern African Studies* 24, no. 4 (December 1998): 698–718.

———. "Sex Lessons from the Past?" *Agenda* 29 (1996): 79–91.

Burrows, E. A *History of Medicine in South Africa up to the End of the Nineteenth Century*. Cape Town: A. A. Balkema, 1958.

Butchard, A. *The Anatomy of Power: European Constructions of the African Body*. London: Zed Books, 1998.

Butler, A. "South Africa's HIV/Aids Policy, 1994–2004: How Can It Be Explained?" *African Affairs* 104, no. 417 (2005): 591–614.

Callaway, H. "Divination and Analogous Phenomena among the Natives of Natal." *Journal of the Royal Anthropological Institute of Great Britain and Ireland* 1 (1871): 163–85.

———. *Nursery Tales, Traditions and Histories of the Zulus*. Westport, CT: Negro Universities Press, 1970. First published 1868.

———. *The Religious System of the Amazulu in Their Own Words*. London, 1870.

Campbell, S. G. "Zulu Witchdoctors and Experiences amongst the Zulus." *Glasgow Medical Journal* (October 1888). Reprint at Killie Campbell Library, Durban.

Campbell, S. S. *Called to Heal: Traditional Healing Meets Modern Medicine in Southern Africa Today*. Johannesburg: Zebra Press, 1998.

Carlson, D. *African Fever: A Study of British Science, Technology, and Politics in West Africa, 1787–1864*. New York: Science History Publications, 1984.

Carton, B. *Blood from Your Children*. Charlottesville: University Press of Virginia, 2000.

Carton, B., J. Laband, and J. Sithole, eds. *Zulu Identities: Being Zulu, Past and Present*. Pietermaritzburg: University of KwaZulu-Natal Press, 2008.

Cartwright, A. P. *Doctors of the Mines. A History of the Mine Medical Officers' Association of South Africa*. Cape Town: Purnell, 1971.

Chase, J. *The Natal Papers: A Reprint of All Notices and Public Documents Connected with That Territory Including a Description of the Country and a History of Events from 1498 to 1843*. Cape Town: C. Struik, 1968.

Chavunduka, G. *Traditional Healers and the Shona Patient*. Gwelo: Mambo Press, 1978.

Clifton Roberts, C. "Witchcraft and Colonial Legislation." *Africa* 8, no. 4 (1935): 488–93.

Cluver, E. H. *Public Health in South Africa*. Johannesburg: Central News Agency, 1959.

———. *Medical and Health Legislation in the Union of South Africa*. Cape Town: Central News Agency, 1949.

Cobbing, J. "The Absent Priesthood: Another Look at the Rhodesian Risings of 1896–1897." *Journal of African History* 18 (1977): 61–84.

———. "A Tainted Well: The Objectives, Historical Fantasies and Working Methods of James Stuart, with Counter-argument." *Journal of Natal and Zulu History* 11 (1988): 115–54.

Colenso, Rev. *Zulu-English Dictionary*. Natal: Vause, Slatter, 1905.

Comaroff, J. *Body of Power, Spirit of Resistance: The Culture and History of a South African People*. Chicago: University of Chicago Press, 1985.

———. "Medicine: Symbol and Ideology." In *The Problem of Medical Knowledge*, edited by P. Wright and A. Treacher. Edinburgh: Edinburgh University Press, 1982.

———. "Medicine and Culture: Some Anthropological Perspectives." *Social Science and Medicine* 12B (1978): 247–54.

———. "The Diseased Heart of Africa: Medicine, Colonialism, and the Black Body." In *Knowledge, Power and Practice: The Anthropology of Medicine and Everyday Life*, edited by S. Lindebaum and M. Lock, 305–29. Berkeley and Los Angeles: University of California Press, 1993.

Comaroff J., and J. L., eds. *Modernity and Its Malcontents: Ritual and Power in Post-Colonial Africa*. Chicago: University of Chicago Press, 1993.

———. *Of Revelation and Revolution*. Vol. 1, *Christianity, Colonialism, and Consciousness in South Africa*. Chicago: University of Chicago Press, 1991.

Cope, T. *Izibongo: Zulu Praise-Poem*. Oxford: Oxford University Press, 1968.

Cornell, S., and D. Hartmann. *Ethnicity and Race: Making Identities in a Changing World*. Thousand Oaks, CA: Pine Forge Press, 1998.

Cory, G. *The Diary of Rev. Francis Owen*. Cape Town: Van Riebeeck Society, 1926.

Crais, C. *White Supremacy and Black Resistance in Pre-industrial South Africa: The Making of the Colonial Order in the Eastern Cape, 1770–1865*. Johannesburg: Witwatersrand University Press, 1992.

Cunningham, A. "The Herbal Medicine Trade." In *South Africa's Informal Economy*, edited by E. Preston-Whyte and C. Rogerson, 196–206. Cape Town: Oxford University Press, 1991.

———. *An Investigation of the Herbal Medicine Trade in Natal/KwaZulu: Report of the Institute of Natural Resources*. Pietermaritzburg: Institute of Natural Resources, University of Natal, 1988.

Curtin, P. *The Image of Africa: British Ideas and Action, 1780–1850*. Madison: University of Wisconsin Press, 1964.

Dauskardt, R. P. A. "Traditional Medicine: Perspectives and Policies in Health Care Development." *Development Southern Africa* 7, no. 3 (1990): 351–58.

———. "Urban Herbalism: The Restructure of Informal Survival in Johannesburg." In *South Africa's Informal Economy*, edited by E. Preston-Whyte and C. Rogerson, 336–44. Cape Town: Oxford University Press, 1991.

Deacon, H. "Racial Segregation and Medical Discourse in Nineteenth-Century Cape Town." *Journal of South African Studies* 22, no. 2 (June 1996): 287–308.

Dhupelia-Mesthrie, U. *From Cane Fields to Freedom*. Cape Town: Kwela, 2001.

Digby, A. "'A Medical El Dorado'? Colonial Medical Incomes and Practise at the Cape." *Social History of Medicine* 7, no. 3 (1995): 463–79.

———. "Self-Medication and the Trade in Medicine within a Multi-ethnic Context: A Case Study of South Africa from the Mid-nineteenth to the Mid-twentieth Centuries." *Social History of Medicine* 18, no. 3 (2005): 439–57.

Doell, E. W. *Doctor Against Witchdoctor*. London: C. Johnson, 1955.

Dohne, J. L. *A Zulu-Kafir Dictionary Etymologically Explained, with Copious Illustrations and Examples, Preceded by an Introduction on the Zulu-Kafir Language*. Cape Town: Farnborough, 1857.

Doke, C., D. Malcolm, J. Sikakana, and B. Vilakazi. *Zulu-English Dictionary.* Johannesburg: Witwatersrand University Press, 1990. First published 1948.

Dubow, S. *Scientific Racism in Modern South Africa.* Cambridge: Cambridge University Press, 1995.

Duminy, A., and B. Guest, eds. *Natal and Zululand from Earliest Times to 1910.* Pietermaritzburg: University of Natal, 1989.

Durban Housing Survey—Multiracial Community. University of Natal, 1952.

du Toit, B. "The Isangoma: An Adaptive Agent among Urban Zulu." *Anthropological Quarterly* 44, no. 2 (1971): 51–66.

Ebr.-Vally, R. *Kala Pani.* Cape Town: Kwela, 2001.

Echenberg, M. *Black Death, White Medicine.* Oxford: Portsmouth, NH: Heinemann, 2002.

E. G. C. "Zulu Customs: An Interview with an Isanusi, or Kafir Witch-Doctor." *Natal Colonist,* June 19, 1879, reprint at the South African Library, Cape Town.

Edwards, S. D. *Some Indigenous South African Views on Illness and Healing.* KwaDlangezwa, RSA: University of Zululand, 1985.

Elphick, R., and R. Shell. "Intergroup Relations: Khoikhoi, Settlers, Slaves and Free Blacks, 1652–1795." In *The Shaping of South African Society, 1652–1820,* edited by R. Elphick and H. Giliomee, chapter 4. Cape Town: Longman, 1979.

Ernst, W., ed. *Plural Medicine, Tradition and Modernity, 1800–2000.* New York: Routledge, 2001.

Etherington, N. "Missionary Doctors and African Healers in Mid-Victorian South Africa." *South African Historical Journal* 19 (1987): 77–91.

———. *Preachers, Peasants, and Politics in Southeast Africa, 1835–1880.* London: Royal Historical Society, 1978.

———. "The 'Shepstone System' in the Colony of Natal and beyond the Borders." In *Natal and Zululand: From Earliest Times to 1910,* edited by A. Duminy and B. Guest, 170–92. Natal: University of Natal Press, 1989.

Evans-Pritchard, E. E. "The Morphology and Function of Magic." *American Anthropologist,* n.s., 31, no. 4 (1929): 619–41.

———. "Witchcraft." *Africa: Journal of the International Institute of African Languages and Cultures* 8, no. 4 (1935): 416–22.

———. *Witchcraft, Oracles and Magic among the Azande.* Oxford: Oxford University Press, 1937.

Falola, T., and Ityavyar, D. *The Political Economy of Health in Africa.* Athens: Ohio University Press, 1992.

Fanon, F. "Medicine and Colonialism." In *A Dying Colonialism,* 121–45. New York: Monthly Review Press, 1965.

———. *Wretched of the Earth.* New York: Grove Weidenfeld, 1963.

Farley, J. Bilharzia. *A History of Imperial Tropical Medicine.* Cambridge: Cambridge University Press, 1991.

Feierman, S. "Change in African Therapeutic Systems." *Social Science and Medicine* 13B (1979): 277–85.

———. "Healing as Social Criticism in the Time of Colonial Conquest." *African Studies* 54 (1995): 72–88.

———. *Peasant Intellectuals: Anthropology and History in Tanzania*. Madison: University of Wisconsin Press, 1990.

———. "Struggles for Control: The Social Roots of Health and Healing in Modern Africa." *African Studies Review* 28, no. 1 (1985): 73–147.

Feierman, S., and J. Janzen, eds. *The Social Basis of Health and Healing in Africa*. Berkeley: University of California Press, 1992.

Fett, S. *Working Cures*. Chapel Hill: University of North Carolina Press, 2002.

Ford, J. *The Role of Trypanosomiasis in African Ecology: A Study of the Tsetse Fly Problem*. Oxford: Clarendon Press, 1971.

Frack, I. *A South African Doctor Looks Backwards and Forward*. Johannesburg: Central News Agency, 1943.

Fry, P. *Spirits of Protest: Spirit-Mediums and the Articulation of Consensus among the Zezuru of Southern Rhodesia (Zimbabwe)*. Cambridge: Cambridge University Press, 1976.

Fuze, M. *The Black People and Whence They Came*. Translated by H. C. Lugg. Durban: Killie Campbell Africana Library, 1979.

Gale, G. W. "Doctor or Witch-Doctor? The Old Medicine and the New." Pamphlet. Johannesburg: South African Institute of Race Relations, 1943.

Gardiner, A. *Narrative of a Journey to The Zoolu Country, in South Africa*. London, 1836.

Gardiner, F., ed. *South African Criminal Law and Procedure*. 2 vols. Cape Town: Juta, 1946.

Gelfand, M., and P. Laidler. *South Africa: Its Medical History, 1652–1898*. Cape Town: C. Struik, 1971.

Gelfand, M. *Christian Nurse and Doctor*. Santon, Aitken family and friends, 1984.

———. *Godly Medicine in Zimbabwe: A History of Its Medicial Missions*. Gweru: Mambo, 1988.

Gerstner, J. "A Preliminary Checklist of Zulu Names of Plants." *Bantu Studies* 12 (1938): 215–36.

Geshiere, P. *The Modernity of Witchcraft*. Charlottesville: University Press of Virginia, 1997.

Gibson, J. Y. *The Story of the Zulus*. Pietermaritzburg: Davis and Sons, 1903.

Gilman, S. *Inscribing the Other*. Lincoln: University of Nebraska Press, 1991.

Gluckman, M. "The Rise of the Zulu Empire." *Scientific American* 202, no. 4 (1969): 157–68.

———. "Zulu Women in Hoe Cultural Ritual." *Bantu Studies* 9 (1935): 255–71.

Gomez, M. *Exchanging Our Country Marks*. Chapel Hill: University of North Carolina Press, 1998.

Good, B. *Medicine, Rationality, and Experience: An Anthropological Perspective*. Cambridge: Cambridge University Press, 1996.

Good, C. *Ethnomedical Systems in Africa*. New York: Guilford Press, 1987.

Gray, N. "Independent Spirits: The Politics of Policing Anti-Witchcraft Movements in Colonial Ghana, 1908–1927." *Journal of Religion in Africa* 35, no. 2 (2005): 139–58.

———. "Witches, Oracles, and Colonial Law: Evolving Anti-Witchcraft Practices in Ghana, 1927–1932." *International Journal of African Historical Studies* 34, no. 2 (2001): 339–63.

Gray, S., ed. *The Natal Papers of John Ross*. Durban: Killie Campbell Africana Library Press, 1992.

Green, E. *AIDS and Sexually Transmitted Disease in Africa: Bridging the Gap between Traditional Healers and Modern Medicine*. Oxford: Westview Press, 1994.

———. *Indigenous Theories of Contagious Disease*. Walnut Creek, CA: AltaMira Press, 1999.

Grossberg, L., C. Nelson, and P. Treichler, eds. *Cultural Studies*. New York: Routledge, 1992.

Gumede, M. V. *Traditional Healers*. Cape Town: Scotaville, 1990.

Gump, J. O. *The Formation of the Zulu Kingdom in South Africa, 1750–1840*. New York: Mellen Press, 1990.

Gussler, J. "Social Change, Ecology, and Spirit Possession among the South African Nguni." In *Religion, Altered States of Consciousness, and Social Change*, edited by E. Bourguignon, 88–126. Columbus: Ohio State University Press, 1973.

Gutherie, M. *Comparative Bantu: An Introduction to the Comparative Linguistics and Prehistory of the Bantu Languages*. 4 vols. Farnborough: Gregg, 1967–71.

Guy, J. *Destruction of the Zulu Kingdom: Civil War in Zululand, 1879–1884*. London: Longman, 1979.

———. "Gender Oppression in Southern Africa's Pre-Capitalist Societies." In *Women and Gender in Southern Africa*, edited by C. Walker, 33–47. Cape Town: David Philip, 1990.

———. *The Maphumulo Uprising*. Pietermaritzburg: University of KwaZulu–Natal Press, 2005.

Haines, R., and G. Buijs, eds. *The Struggle for Social and Economic Space: Urbanisation in Twentieth Century South Africa*. Durban: Institute for Social and Economic Research, University of Durban-Westville, 1985.

Hale, Frederick, ed. *Norwegian Missionaries in Natal and Zululand: Selected Correspondence, 1844–1900*. Cape Town: Van Riebeeck Society, 1997.

Hall, M. *Farmers, Kings, and Traders: The People of Southern Africa, 200–1860*. Chicago: University of Chicago Press, 1990.

Hamilton, C., ed. *The Mfecane Aftermath*. Johannesburg: Witwatersrand University Press, 1995.

———. *Terrific Majesty: The Powers of Shaka Zulu and the Limits of Historical Invention*. Johannesburg: Witwatersrand University Press, 1998.

Hammond-Tooke, W. D. *Rituals and Medicines: Indigenous Healing in South Africa*. Johannesburg: Ad. Porter, 1989.

Hanretta, S. "Women, Marginality and the Zulu State: Women's Institutions and Power in the Early Nineteenth Century." *Journal of African History* 39, no. 3 (1998): 389–415.

Harries, P. *Work, Culture, and Identity: Migrant Labourers in Mozambique and South Africa, c. 1860–1910*. London: James Currey, 1994.

Hartwig, G. W., and K. D. Patterson. *Disease in African History: An Introductory Survey and Case Studies*. Durham, NC: Duke University Press, 1978.

Hattersley, A., ed. *Later Annals of Natal*. London, Green, 1938.

———. *More Annals of Natal*. Pietermaritzburg: Shuter and Shooter, 1936.

———. *The Natalians: Further Annals of Natal*. Pietermaritzburg: Shuter and Shooter, 1940.

Heelas, P., S. Lash, and P. Morris. *Detraditionalization: Critical Reflections on Authority and Identity at a Time of Uncertainty*. Oxford: Blackwell, 1996.

Henderson, W. P. M. *Durban: Fifty Years of Municipal History*. Durban: Robinson, 1904.

Hewat, M. *Bantu Folklore: Medical and General*. Westport, CT: Negro University Press, 1979; first impression, 1906.

Hitchins, R. L., ed. *Natal Statutes 1845–1899*. 3 vols. Pietermaritzburg, 1901.

Hobsbawm, E., and T. Ranger, eds. *The Invention of Tradition*. Cambridge: Cambridge University Press, 1983.

Holden, W. C. *British Rule in South Africa*. London: Wesleyan Conference Office, 1879.

Hund, J., ed. *Witchcraft Violence and the Law in South Africa*. Pretoria: Protea Book House, 2003.

Hunt, N. R. *A Colonial Lexicon of Birth Ritual, Medicalization, and Mobility in the Congo*. Durham, NC: Duke University Press, 1999.

Hunter, M. *Reaction to Conquest: Effects of Contact with Europeans on the Pondo of South Africa*. London, Oxford University Press, 1936.

Hutchings, A. *Zulu Medicinal Plants: An inventory*. Pietermaritzburg: University of Natal Press, 1996.

Hutnyk, J. *Critique of Exotica Music, Politics and the Culture Industry*. London: Pluto Press, 2000.

Iliffe, J. *The African AIDS Epidemic*. Athens: Ohio University Press, 2006.

———. *East African Doctors: A History of the Modern Profession*. Cambridge: Cambridge University Press, 1998.

Isaacs, N. *Travels and Adventures in Eastern Africa*. Cape Town: C. Struik, 1970.

Jackson, C. G. "The Medical Man in Natal and Zululand." *South African Journal of Science* 5 (1919): 251–60.

Jacobson-Widding, A., and D. Westerlund, eds. *Culture, Experience and Pluralism: Essays on African Ideas of Illness and Healing*. Stockholm: Uppsala University, 1989.

Jain, P. *Indians in South Africa*. Delhi: Kalinga Publications, 1999.

Jain, R. *Indian Communities Abroad*. New Delhi: Manohar, 1993.

Janzen, J. *Ngoma: Discourses in Healing*. Berkeley and Los Angeles: University of California Press, 1992.

———. *The Quest for Therapy in Lower Zaire*. Berkeley and Los Angeles: University of California Press, 1978.

Janzen, J., and E. Green. "Continuity, Change, and Challenge in African Medicine." In *Medicine across Cultures: History and Practice of Medicine in Non-Western Cultures*, edited by H. Selin, 1–26. Boston: Kluwer Academic Publishers, 2003.

Jenkins, R. *Rethinking Ethnicity*. Thousand Oaks, CA: Sage, 1997.

Jenkinson, T. *Amazulu: The Zulus, Their Past History, Manners, Customs, and Language: Natal 1873–1879*. London, 1882.

Jochelson, K. J. *The Colour of Disease: Syphilis and Racism in South Africa*. Basingstoke: Palgrave, 2001.

———. "Sexually Transmitted Diseases in Nineteenth and Twentieth Century South Africa." In *Histories of Sexually Transmitted Diseases and HIV/AIDS in Sub-Saharan Africa*, edited by P. Setel et al., 217–44. Westport, CT: Greenwood, 1999.

Johnson, T., and C. Sargent. *Medical Anthropology: A Handbook of Theory and Method*. New York: Praeger, 1990.

Junod, H. "The Theory of Witchcraft amongst South African Natives." In *Report of the South African Association for the Advancement of Science* (1907): 230–41.

Kelley, R. D. G. "People in Me." *Colorlines* 1, no. 3 (1999): 5–7.

Kertzer, D. *Ritual, Politics, and Power*. New Haven, CT: Yale Unversity Press, 1988.

Kiple, K., ed. *Cambridge World History of Human Disease*. New York: Cambridge University Press, 1993.

Kirby, P., ed. *Andrew Smith and Natal: Documents Relating to the Early History of that Province*. Cape Town: Van Riebeeck Society, 1955.

Kline, B. *Genesis of Apartheid: British African Policy in the Colony of Natal, 1845–1893*. Lanham: University Press of America, 1988.

Klugman, B. "The Politics of Contraception in South Africa." *Women's Studies International Forum* 13, no. 3 (1990): 261–71.

Knight, I. *The Anatomy of the Zulu Army*. London: Greenhill Books, 1995.

Kodesh, N. "Renovating Tradition: The Discourse of Succession in Colonial Buganda." *International Journal of African Historical Studies* 34, no. 3 (2001): 511–41.

Kohler, M. *The Izangoma Diviners*. Edited and translated by N. J. van Warmelo. Ethnological Publications 9. Pretoria: Government Printers, 1941.

Kramer, P. *Listening to Prozac*. New York: Viking Press, 1993.

Krige, E. *The Social System of the Zulus*. London, 1936.

Krige, J. D. "The Social Function of Witchcraft." *Theoria* 1 (1947): 8–21.

Kunene, M. *Anthem of the Decades: A Zulu Epic*. London: Heinemann, 1986.

———. *Emperor Shaka the Great: A Zulu Epic*. London: Heinemann, 1979.

Kunnie, J. "Indigenous African Churches and Religio-Cultural Liberation: The Practice of Traditional Healing." Paper presented at the African Studies Association, Seattle, Washington, 20–23 November 1992.

Kuper, H. *Indian People in Natal.* Pietermaritzburg: Natal University Press, 1960.
———. "Medicine among the South Eastern Bantu." *The Leech,* November 1936.
———. "'Strangers' in Plural Societies." In *Pluralism in Africa,* edited by L. Kuper and M. G. Smith, 247–82. Berkeley and Los Angeles: University of California Press, 1969.
Kuper, L., H. Watts, and R. Davies. *Durban: A Study in Racial Ecology.* London: J. Cape, 1958.
Kuper, P., and F. Meer. "Indian Elites in Natal, South Africa." In *University of Natal Institute for Social Research Proceedings* (July 1956).
Laband, J. *Rope of Sand: The Rise and Fall of the Zulu Kingdom in the Nineteenth Century.* Johannesburg: Jonathan Ball, 1995.
Lambert, J. *Betrayed Trust: Africans and the State in Colonial Natal.* Pietermaritzburg: University of Natal Press, 1995.
Lan, D. *Guns and Rain.* London: J. Currey, 1987.
Landau, P. "Explaining Surgical Evangelism in Colonial Southern Africa: Teeth, Pain, and Faith." *Journal of African History* 37, no. 2 (1996): 261–81.
Last, M., and G. Chavunduka, eds. *The Professionalisation of African Medicine.* Manchester: Manchester University Press, 1986.
Lawson, T. *Religions of Africa: Traditions in Transformation.* San Francisco: Harper anad Row, 1984.
LeClerc-Madlala, S. "Virginity Testing: Managing Sexuality in a Maturing HIV/ AIDS Epidemic." *Medical Anthropology Quarterly* 15, no. 4 (2001): 533–52.
Lee, S. G. "Spirit Possession among the Zulu." In *Spirit Mediumship and Society in Africa,* edited by J. Beattie and J. Middleton, 128–56. London: Routledge, 1969.
Leslie, C. "Medical Pluralism and Legitimation in the Indian and Chinese Medical Systems." In *Culture, Disease and Healing: Studies in Medical Anthropology,* edited by D. Landy, 511–17. New York: Macmillan, 1977.
Leslie, D. *Among the Zulus and Amatongas.* New York: Negro Universities Press, 1969; First published Glasglow: W. Gilchrist, 1875.
Le Suer, D. "Historical Perspective of the Malaria Problem in Natal with emphasis on the Period 1928 to 1932." *South African Journal of Science* 89, no. 5 (1993): 232–39.
Livingston, D. *Missionary Travels and Researches in South Africa.* Philadelphia, 1858.
Louw, D. A., and Pretorius, E. "The Traditional Healer in a Multicultural Society: The South African Experience." In *Spirit versus Scalpel: Traditional Healing and Modern Psychotherapy,* edited by L. L. Adler and B. R. Mukherji, 41–58. London: Bergin and Garvey, 1995.
Luedke, T., and H. West, eds. *Borders and Healers.* Bloomington: Indiana University Press, 2006.
Lugg, H. C. *Life under a Zulu Shield.* Pietermaritzburg: Shuter and Shooter, 1975.
———. *A Natal Family Looks Back.* Durban: T. W. Greggs, 1970.

Lyons, M. *The Colonial Disease: A Social History of Sleeping Sickness in Northern Zaire, 1900–1940.* Cambridge: Cambridge University Press, 1992.

MacKinnon, A. "'Of Oxford Bags and Twirling Canes': The State, Popular Responses, and Zulu Antimalaria Assistants in the Early-Twentieth-Century Zululand Malaria Campaigns." *Radical History Review* 80 (Spring 2001): 76–100.

Macleod, R., and Lewis, M., eds. *Disease, Medicine, and Empire: Perspectives on Western Medicine and the Experience of European Expansion.* New York: Routledge, 1988.

Magubane, Z. *Bringing the Empire Home: Race, Class, and Gender in Britain and Colonial South Africa.* Chicago: University of Chicago Press, 2004.

Mahoney, M., and Parle, J. "An Ambiguous Sexual Revolution? Sexual Change and Intra-generational Conflict in Colonial Natal." *South African Historical Journal* no. 50 (2004): 134–51.

Maier, D. "Nineteenth-Century Asante Medical Practices." *Comparative Studies in Society and History* 21 (1979): 63–81.

Mamdani, M. *Citizen and Subject: Contemporary Africa and the Legacy of Late Colonialism.* Princeton, NJ: Princeton University Press, 1996.

Mander, M. *Marketing of Indigenous Medical Plants in South Africa.* Rome: FAO United Nations Press, 1998.

Maphalala, S. J. *Aspects of Zulu Rural Life during the Nineteenth Century.* Empangeni: University of Zululand, 1985.

Marks, S. *The Ambiguities of Dependence in South Africa.* Baltimore: Johns Hopkins University Press, 1986.

———. *Divided Sisterhood: Race, Class, and Gender in the South African Nursing Profession.* New York: St. Martin's Press, 1994.

Marks, S., and A. Atmore, eds. *Economy and Society in Pre-industrial South Africa.* London: Longman, 1980.

Marks, S., and R. Rathbone. *Industrialisation and Social Change in South Africa.* London: Longman, 1985.

Martin, E. "The Egg and the Sperm: How Science Has Constructed a Romance Based on Stereotypical Male-Female Roles." *Signs* 16, no. 3 (Spring 1991): 485–501.

Mayat, Z., ed. *Indian Delights.* Durban: Women's Cultural Group, 1998. First published 1961.

Maylam, P., and I. Edwards. *The People's City.* Pietermaritzburg: University of Natal Press, 1996.

Mbali, M. "A Medical History 'From Below': A Critical Review of New Literature on Changes in African Culture in South Africa and STD and AIDS EPI." *Journal of Natal and Zulu History* 21 (2003): 77–93.

Mbiti, J. *African Religion and Philosophy.* London: 1969.

McClintock, A. *Imperial Leather.* New York: Routledge, 1995.

McCord, J. "Medical Missionary Work in Africa." *Christianity and the Native of South Africa: A Yearbook of South African Missions.* South Africa, 1928.

———. *My Patients Were Zulus*. New York: Rinehart, 1951.

———. "Native Witch Doctors and Healers." *South African Medical Record* 24 (1926).

———. "Some Sidelights on the House that Jim Built." Killie Campbell, Pamphlet 362 MACC, 1929.

———. "A Zulu Witch Doctor and Medicine Man." *South African Medical Record* 16, no. 8 (April 27, 1918): 116–22.

McCord, M. *The Calling of Katie Makanya*. Cape Town: David Philip, 1995.

McCracken, D., and P. McCracken. *Natal: The Garden Colony*. Sandton: Frandsen, 1990.

Medley Wood, J. "Native Herbs: Medicinal and Otherwise." *Natal Almanac* 1898. Natal, 1898.

Meer, F. *Portrait of Indian South Africans*. Durban: Avon House, 1969. First published 1843.

Moffat, R. *Missionary Labours and Scenes in Southern Africa*. New York: Johnson Report Corporation, 1969.

Mofolo, T. *Chaka the Zulu*. London: Oxford University Press, 1949.

Molema, S. M. *The Bantu — Past and Present*. Edinburgh: Green and Sons, 1920.

Morrell, R, ed. *Political Economy and Identities in KwaZulu-Natal: Historical and Social Perspectives*. Durban: Indicator Press, 1996.

Mudimbe,V. Y. *The Invention of Africa: Gnosis, Philosophy, and the Order of Knowledge*. Bloomington: Indiana University Press, 1988.

Nesvag, S. "Street Trading from Apartheid to Post-Apartheid: More Birds in the Cornfield?" *International Journal of Sociology and Social Policy* 21, nos. 3/4 (2000): 34–63.

Niehaus, I. *Witchcraft, Power, and Politics: Exploring the Occult in the South African Lowveld*. Cape Town: David Philip, 2001.

Ngubane, H. "Aspects of Clinical Practice and Traditional Organization of Indigenous Healers in South Africa." *Social Science and Medicine* 15, no. 3 (1981): 361–65.

———. *Body and Mind in Zulu Medicine: An Ethnography of Health and Disease in Nyuswa-Zulu Thought and Practice*. London: Academic Press, 1977.

Nuttal, Sarah, and Carli Coetzee, eds. *Negotiating the Past: The Making of Memory in South Africa*. London: Oxford University Press, 1998.

Olumwullah, O. *Dis-ease in the Colonial State: Medicine, Society and Social Change among the AbaNyole or Western Kenya*. Westport, CT: Greenwood Press, 2002.

Omer-Cooper, J. D. *The Zulu Aftermath: A Nineteenth-Century Revolution in Bantu Africa*. Evanston: Northwestern University Press, 1966.

Orde Browne, G. St. J. "Witchcraft and British Colonial Law." *Africa, Journal of the International Institute of African Languages and Cultures* 8, no. 4 (1935): 481–87.

Oudshoorn, N. *Beyond the Natural Body: An Archeology of Sex Hormones*. London: Routledge, 1994.

Packard, R. *White Plague, Black Labor: Tuberculosis and the Political Economy of Health and Disease in South Africa.* Berkeley and Los Angeles: University of California Press, 1989.

Panikar, S. *The Politics of Racialism.* Agra: Shiva Lal Agarwala, 1962.

Pappe, L. *Florae capensis medicae prodromus, or, An enumeration of South African indigenous plants : used as remedies by the colonists of the Cape of Good Hope.* Cape Town, 1850.

Parle, J. *States of Mind: Searching for Mental Health in Natal and Zululand, 1868–1918.* Pietermaritzburg: University of KwaZulu-Natal Press, 2007.

——. "Witchcraft or Madness? The *Amandiki* of Zululand, 1894–1914." *Journal of Southern African Studies* 29, no. 1 (March 2003): 105–32.

Parnell, S. "Creating Racial Privilege: The Origins of South African Public Health and Town Planning Legislation." *Journal of Southern African Studies* 19, no. 3 (September 1993): 471–88.

Peires, J., ed. *Before and After Shaka: Papers in Nguni History.* Institute of Social and Economic Research, Rhodes University: Grahamstown, 1981.

——. *The Dead Will Arise.* Johannesburg: Ravan Press, 1989.

Phillips, H. *Black October: The Impact of the Spanish Influenza Epidemic of 1918 on South Africa.* Cape Town: University of Cape Town, 1991.

Pieterse, J. *White on Black: Images of Africa and Blacks in Western Popular Culture.* New Haven, CT: Yale University Press, 1992.

Posey, D., and G. Dutfield. *Beyond Intellectual Property.* Ottawa: IDRC, 1996.

Prashad, V. "Afro-Dalits of the Earth, Unite!" *African Studies Review* 43, no. 1 (2000): 189–201.

——. *Everybody Was Kungfu Fighting.* Boston: Beacon Hill Press, 2001.

Ranger, T. "The Invention of Tradition Revisted: The Case of Colonial Africa." In *Legitimacy and the State in Twentieth-Century Africa*, edited by T. Ranger and O. Vaughn, 62–111. New York: Palgrave, 1993.

Ranger, T., and P. Slack, eds. *Epidemics and Ideas: Essays on the Historical Perception of Pestilence.* New York: Cambridge University Press, 1992.

Reader's Digest Illustrated History of South Africa. Cape Town: Reader's Digest Association, 1995.

Rees, H. "The Abortion Debate in South Africa." *Critical Health* 34 (June 1991): 20–26.

Reid, W. V. "Bioprospecting: A Force for Sustainable Development." *Environmental Science and Technology* 27, no. 9 (1993): 1730–32.

Reyher, R. H. *Zulu Woman: The Life Story of Christina Sibiya.* New York: Feminist Press at CUNY, 1999. First published 1948.

Rheinallt Jones, J. D., and A. L. Saffery. "Social and Economic Conditions of Native Life in the Union of South Africa: Findings of the Native Economic Commission, 1930–1932, Collated and Summarized." *Bantu Studies* 7, no. 3 (1933).

Rogerson, C. M. "Shisha Nyama: The Rise and Fall of the Native Eating House Trade in Johannesburg." *Social Dynamics* 14, no. 1 (1988): 20–33.

Russell, G. *History of Old Durban. Reminiscences of an Emigrant of 1850*. Durban: Davis and Sons, 1899.

Russett, C. *Sexual Science: The Victorian Construction of Womanhood*. Cambridge, MA: Harvard University Press, 1989.

Samuelson, R. *King Cetewayo Zulu Dictionary*. Durban: Commercial Printing, 1923.

——. *Long, Long Ago*. Durban: Knox, 1929.

Saunders, C. *Historical Dictionary of South Africa*. Lanham, MD: Scarecrow Press, 1983.

Schimlek, F. *Medicine versus Witchcraft*. Natal: Mariannhill Mission Press, 1950.

Scorgie, F. "Virginity Testing and the Politics of Sexual Responsibility: Implications for AIDS Intervention."*African Studies* 61, no. 1 (2002): 55–75.

Shapiro, K. "Doctors or Medical Aids—The Debate over the Training of Black Medical Personnel for the Rural Black Population in South Africa in the 1920s and 1930s." *Journal of Southern African Studies* 13, no. 2 (1987): 234–55.

Shaw, R. "The Politics and the Diviner: Divination and the Consumption of Power in Sierra Leone." *Journal of Religion in Africa* 26 (February 1996): 30–55.

Shils, E. *Tradition*. Chicago: University of Chicago Press, 1981.

Shiva, V. *Biopiracy: The Plunder of Nature and Knowledge*. Boston: South End, 1997.

Shooter, Rev. J. *The Kafirs of Natal and the Zulu Country*. London, 1857.

Sibeko, V. "Imilingo nemithi yokwelapha phakathi kwabantu" (Magic and healing medicines of the people). *Natal Native Teacher's Journal* 11, no. 3 (April 1932), and 11, no. 4 (July 1932).

Simon, L. *Inyanga: Sarah Mashele's Story*. Johannesburg: Justified, 1993.

Simons, R., and C. Hughs, eds. *The Culture-Bound Syndromes*. Boston: D. Reidel, 1985.

Singer, P., ed. *Traditional Healing: New Science or New Colonialism?* (Essays in Critique of Medical Anthropology). New York: Conch Magazine, 1977.

Smith, A. *A Contribution to South African Materia Medica Chiefly from Plants in Use among the Natives*. Lovedale, RSA, 1888.

Sookdeo, A. "The Transformation of Ethnic Identities: The Case of 'Coloured' and Indian Africans." *Journal of Ethnic Studies* 15, no. 4 (1988): 69–83.

South African Archival Records. 4 vols. Cape Town: The Branch, 1962, 1963.

South African Law Reports: Decisions of the Supreme Courts of South Africa, the High Court of Southern Rhodesia, the High Court of South-West Africa. 3 vols. Cape Town: Juta, 1947.

Spiegel, A. D., and P. A. McAllister, eds. *Tradition and Transition in Southern Africa*. London: Transaction, 1991.

Speight, W. L. "Witchcraft and Witchdoctors." *Chamber's Journal* (1929): 835–55.

Spohr, O. H. *The Natal Diaries of Dr. W. H. I. Bleek, 1855–1856*. Cape Town: A. A. Balkema, 1965.

Stafford, W. G. *Principles of Native Law*. Pietermaritzburg: Shuter and Shooter, 1955.

Starr, P. *The Social Transformation of American Medicine*. New York: Basic Books, 1982.

Stoler, A. *Race and the Education of Desire: Foucault's* History of Sexuality *and the Colonial Order of Things.* Durham, NC: Duke University Press, 1995.

Stuart, J., ed. *The Diary of Henry Francis Fynn.* Pietermaritzburg: Shuter and Shooter, 1950.

Sundkler, B. *Bantu Prophets in South Africa.* London: Oxford University Press, 1961.

Swanson, M. "The Sanitation Syndrome: Bubonic Plague and Urban Native Policy in the Cape Colony, 1900–1909." *Journal of African History* 18, no. 3 (1979): 387–410.

Takaki, R. *A Different Mirror.* Boston: Little, Brown, 1993.

Tomlinson, J. *Cultural Imperialism: A Critical Introduction.* Baltimore: Johns Hopkins University Press, 1991.

Tonkin, E. "Consulting Ku Jlople: Some Histories of Oracles in West Africa." *Journal of the Royal Anthropological Institute* 10 (2004): 539–60.

Tuhiwai Smith, L. *Decolonizing Methodologies.* London: Zed Books, 1999.

Turner, V. *The Drums of Affliction.* Oxford: Clarendon Press, 1968.

Turrell, R. "Muti Ritual Murder in Natal: From Chiefs to Commoners (1900–1930)." *South African Historical Journal* 44 (2001): 21–39.

Tyler, J., Rev. *Forty Years among the Zulus.* Boston, 1891.

Vahed, G. "Control and Repression: The Plight of Indian Hawkers and Flower Sellers in Durban, 1910–1948." *International Journal of African Historical Studies* 32, no. 1 (1999): 19–48.

Vail, L., ed. *The Creation of Tribalism in Southern Africa.* Berkeley and Los Angeles: University of California Press, 1991.

Van Dijk, R., R. Reis, and M. Spieranburg, eds. *The Quest for Fruition through Ngoma: The Political Aspects of Healing in Southern Africa.* Athens: Ohio University Press, 2000.

Van Wyk, B., B. Van Oudtshoorn, and N. Gericke. *Medical Plants of South Africa.* Pretoria: Briza, 1997.

———. *People's Plants,* Pretoria: Briza Publications, 2000.

Vansina, Jan. *Oral Tradition as History.* Madison: University of Wisconsin Press, 1985.

———. *Paths in the Rainforest.* Madison: University of Wisconsin Press, 1990.

Vaughan, M. *Curing Their Ills: Colonial Power and African Illness.* Stanford: Stanford University Press, 1991.

———. "Healing and Curing: Issues in the Social History and Anthropology of Medicine in Africa." *Social History of Medicine* 7, no. 2 (1994): 283–95.

Veritas. "The Kafir Labour Question, as it Affects the Colony of Natal." 1851, housed at the South African Library, Cape Town.

Vijn, C. *Cetshwayo's Dutchman.* New York: Negro Universities Press, 1968.

Vilakazi, A. *Zulu Transformations: A Study of the Dynamics of Social Change.* Pietermaritzburg: University of Natal Press, 1962.

Waite, G. *A History of Traditional Medicine and Health Care in Precolonial East Central Africa.* New York: Edwin Mellen Press, 1992.

Warwick, A. "Immunities of Empire: Race, Disease, and the New Tropical Medicine, 1900–1920." *Bulletin of the History of Medicine* 70 (1996): 94–118.

Watt, J. "The Native Medicine Man." *South African Nursing Record* (July 1931): 262–66.

Watt, J., and M. G. Breyer-Brandwijk. *The Medical and Poisonous Plants of Southern and Eastern Africa: Being an Account of the Medical and Other Uses, Chemical Composition, Pharmacological Effects, and Toxicology in Man and Animals*. Edinburgh, 1962. First published 1932.

Webb, C. B., and J. B. Wright, eds. *The James Stuart Archive of Recorded Oral Evidence Relating to the History of the Zulu and Neighbouring Peoples*. 5 vols. Pietermaritzburg: University of Natal Press, 1976, 1979, 1982, 1986, 2001.

——. *A Zulu King Speaks: Statements Made by Cetshwayo kaMpande on the History and Customs of His People*. 1880. Reprint, Pietermaritzburg: University of Natal Press, 1987.

Wellings et al. "Sexual Behaviour in Context: A Global Perspective." *The Lancet*, November 11, 2006, 368, 9548, 1706–28.

Welsh, D. *The Roots of Segregation: Native Policy in Colonial Natal, 1845–1910*. Cape Town: Oxford University Press, 1975.

White, L. "'They Could Make Their Victims Dull': Genders and Genres, Fantasies and Cures in Colonial Southern Uganda." *American Historical Review* 100, no. 5 (1995): 1379–1402.

Wilson (Hunter), M. "Witch Beliefs and Social Structure." *American Journal of Sociology* 56, no. 4 (1951): 307–13.

Wilson (Hunter), M., and L. Thompson, eds. *The Oxford History of South Africa*. Vol. 1, *South Africa to 1870*. New York: Oxford University Press, 1969.

Worger, B. "Clothing Dry Bones: The Myth of Shaka." *Journal of African Studies* 6, no. 3 (1979): 144–58.

Wylie, D. *Starving on a Full Stomach*. Charlottesville: University of Virginia Press, 2001.

Yoder, S., ed. *African Health and Healing Systems: Proceedings of a Symposium*. Los Angeles: Crossroads Press, 1982.

UNPUBLISHED SOURCES

Augusto, G. "Knowing Differently, Innovating Together? An Exploratory Case Study of Trans-Epistemic Interaction in a South African Bioprospecting Program." PhD diss., George Washington University, 2004.

Balfour-Cunninham, A. "Imithi Isizulu: The Traditional Medicine Trade in Natal/KwaZulu." Master's thesis, University of Natal, Durban, 1993.

Bradford, H. "'Her Body, Her Life': 150 Years of Abortion in South Africa." Unpublished paper, 1991.

Burns, C. "Brought to Bed: Bridgeman Memorial Hospital and the History of Black Women's Experiences of Childbirth in Johannesburg, 1928–1965." PhD diss., Northwestern University, 1996.

Dauskardt, R. P. A. "The Evolution of Health Systems in a Developing World Metropolis: Urban Herbalism on the Witswatersrand." Master's thesis, University of the Witwatersrand, Johannesburg, 1994.

Dube, S. "Medicine in the Economic Life of the Zulu Village." Bachelor's thesis, UNISA, 1975.

Devenish, A. "Negotiating Healing: The Professionalisation of Traditional Healers in KwaZulu-Natal between 1985 and 2003." Master's thesis, University of Natal, Durban, 2003.

Fenyves, K. E. "South African Traditional Healer's Organizations in the Context of Traditionalism and Modernity." Master's thesis, sociology, University of Witwatersrand.

Flint, K. "Negotiating a Hybrid Medical Culture: African Healers in Southeastern Africa from the 1820s to 1940s." PhD diss., University of California, Los Angeles, 2001.

Kaplan, A. "The Marketing of Branded Medicine to the Zulu Consumer." University of Natal, Durban, 1977.

La Hausse, P. "The Struggle for the City: Alcohol, the Ematsheni, and Popular Culture in Durban, 1902–1936." Master's thesis, University of Cape Town, 1984.

Louw, J. K. "Witchcraft among the Southern Bantu: Facts, Problems, Policies." PhD diss., Yale University, 1941.

Mael, R. "The Problem of Political Integration in the Zulu Empire." PhD diss., University of California, Los Angeles, 1974.

Mahoney, M. "Between the Zulu King and the Great White Chief: Political Culture in a Natal Chiefdom, 1879–1906." PhD diss., University of California, Los Angeles, 1998.

Monamodi, I. S. "Medical Doctors under Segregation and Apartheid: A Sociological Analysis of Professionalization among Doctors in South Africa, 1900–1980." PhD diss., Indiana University, 1996.

Nesvag, S. "D'Urbanised Tradition: The Restructuring and Development of the *Muthi* Trade in Durban." Master's thesis, University of Natal, Durban, 1999.

Nieuwenhuijsen, J. W. "Diviners and Their Ancestor Spirits: A Study of the Izangoma among the Nyuswa in Natal, South Africa." PhD diss., University of Amsterdam, 1974.

Parle, J. "States of Mind: Mental Illness and the Quest for Mental Health in Natal and Zululand, 1868–1918." PhD diss., University of KwaZulu-Natal, 2004.

Parle, J., and F. Scorgie. "Bewitching Zulu Women: Umhayiso, Gender and Witchcraft in Natal." Unpublished paper presented at the African Studies Association, Houston, Texas, November 2001.

Seedat, Z. "The Zanzibaris in Durban: A Social Anthropological Study of the Muslim Descendants of African Freed Slaves Living in the Indian Area of Chatsworth." Master's thesis in African Studies, University of Natal, Durban, 1973.

Swanson, M. W. "The Rise of Multi-Racial Durban: History and Race Policy, 1893–1930." PhD diss., Harvard University, 1964.

Zondi, S. "African Demand and Missionary Charity: The Development of Mission Health Services in Kwazulu to 1919." Unpublished paper presented at the University of KwaZulu-Natal African Studies Seminar, 1999.

Buthezi, A. January 28, 1998, Hlabisa (translator Mbongi Mavunda).

Cele, P. December 9, 1998, Durban.

Eagles, P. (Medicines Control Council). July 15, 2002, Cape Town.

Folb, P. (SATMERG). July 24, 2002, Cape Town.

Govender, L. December 17, 1998, Durban.

Hlabisa. January 27, 1998, Hlabisa.

Jele. January 27, 1998, Hlabisa (translator Mbongi Mavunda).

Joanna. January 28, 1998, Hlabisa (translator Mbongi Mavunda).

Johnson, D. December 9, 1998, Durban.

Koonin, S. (SATMERG). July 12, 2002, Cape Town.

Kubekeli, P. July 19, 2002, Cape Town.

Madikane, E. (SATMERG). July 20, 2002, Cape Town.

Manqele. January 29, 1998, Hlabisa (translator Mbongi Mavunda).

Matsibisi, G. (Medical Research Council). July 18, 2002.

Mbuyazi, J. November 29, 1998, Mthubathuba. (Interviewed with Dingane Mthetwa.)

Mchunu. January 30, 1998, Hlabisa (translator Mbongi Mavunda).

Mhlaba. January 29, 1998, Hlabisa (translator Mbongi Mavunda).

Mhlongo. January 28, 1998, Hlabisa (translator Mbongi Mavunda).

Mhlongo, H. January 29, 1998, Hlabisa (translator Mbongi Mavunda).

Mkhize, V. October 26, 1998, and June 24, 2002, Durban.

Mpongose, Rev. October 30, 1998, Durban.

Mthemjwa, M. October 16, 1998, Durban.

Naidoo, L. March 19, 1998; April 1, 1998; May 8, 1998; July 29, 1998; August 20, 1998; August 28, 1998, all Pietermaritzburg except August 20, Howick Falls.

Naiken, A. June 26, 2002, Durban.

Naiker, V. June 2002, Durban.

Ntuli, Q. December 10, 1998, and June 26, 2002, Durban.

Ntutela, S. (SATMERG). July 20, 2002, Cape Town.

Nxumalo, J. January 29, 1998, Hlabisa.

Pillay, S. June 20, 2002, Newcastle.

Pillay, T. Ladysmith, June 20, 2002.

Soofie, F. June 28, 2002, Durban.

Victoria Market Woman. December 23, 1998, Durban.

Williams, G. August 2, 2002, Cape Town.

Xaba, I. January 26, 1998, Hlabisa (translator Mbongi Mavunda).

Index

Page references in italics denote illustrations.